Foundations of Structural Geology

D EDITION

c, PhD, FGS

Reader in Geology
University of Keele, UK

Blackie

Glasgow and London

Published in the USA by
Chapman and Hall
New York

Blackie & Son Ltd
Bishopbriggs, Glasgow G64 2NZ

7 Leicester Place London WC2H 7BP

Published in the USA by
Chapman and Hall
a division of Routledge, Chapman and Hall, Inc.
29 West 35th Street, New York, NY 10001-2291

British Library Cataloguing in Publication Data

Park, R.G. (R. Graham)
 Foundations of structural geology.—2nd ed.
 1. Geological structure
 I. Title
 551.8

 ISBN 0-216-92491-X
 ISBN 0-216-92492-8 Pbk

For the USA International Standard book numbers are
 0-412-01931-0
 0-412-01941-8 (Pbk.)

To many generations of students who have unwittingly provided the inspiration for this book, and to my wife, for typing and support.

Phototypesetting by Thomson Press (India) Limited, New Delhi

Printed in Great Britain by Scotprint Ltd

Preface

In the Preface to the first edition of this book, published in 1983, I explained my reasons for writing the book as follows.

'There are already a number of excellent books covering the various aspects of Structural Geology. Among these are works by Hobbs, Means and Williams, Jaeger and Cook, Price, Ramsay, and Turner and Weiss, all of which I have used extensively in preparing this book and have listed therein as further reading. However, these textbooks are rather advanced for many students commencing the study of geology, and for many years I have been aware of the lack of a suitable elementary book which I could recommend to beginners. My purpose in writing this book, therefore, was to supplement existing textbooks by providing an introduction to the subject which will convey enough information over the whole field of structural geology to stimulate the reader's interest and encourage further study of more advanced textbooks and scientific papers.'

In the intervening six years since these words were written, the demand for a short, inexpensive, and reasonably comprehensive *elementary* textbook has continued to be just as great.

The revised edition contains a number of completely new sections. Extensional and strike-slip tectonics, and terrane accretion, are dealt with for the first time, reflecting the tremendous growth of interest in these topics among structural geologists over the last few years. I have responded to pleas from Scandinavian colleagues to incorporate a section on stereograms (which forms the Appendix) and an introductory section on elementary geometry and basic concepts, so that the book can be used, if necessary, as the sole introductory text on structural geology. The new introductory section also contains definitions and explanations of a number of important stratigraphic and structural concepts that need to be understood in the early stages of learning structural geology from geological maps. It must be stressed, however, that geological map interpretation must be carried out as a complement to any theoretical treatment of the subject, in order to appreciate structure in three dimensions.

I have also made numerous corrections and improvements to the text and also to the illustrations, many of which have been replaced. In doing this, I have incorporated many helpful suggestions from colleagues and reviewers. In particular, I should like to thank Alan Wright of Birmingham University, who supplied a most helpful and detailed list of suggested improvements. I wish to re-iterate my indebtedness to Paula Haselock, Nick Kusznir, and Rob Strachan (all at that time at Keele), and to two anonymous reviewers who read the draft of the first edition and made many useful suggestions for its improvement. I am especially grateful to Rob Standley, then of City of London Polytechnic, for his meticulous checking of the original manuscript and for a host of valuable suggestions. Many of the original diagrams were drawn by Paula Haselock, whose willing and cheerful help made the task of writing the book much easier.

Finally I wish to make it clear that I have ignored many pieces of good advice in relation to both the first and second editions, usually because of my overriding desire to make the book as short as possible, and that any remaining deficiencies are entirely my own responsibility.

RGP

Contents

Part 3: Geotectonics—major Earth structure

Introduction

Meaning and scope of structural geology. It is easier to give examples of geological structures than to define them. The word 'structure' means 'that which is built or constructed'. Structural geologists use the word to signify something that has been produced by deformation; that is, by the action of forces on and within the Earth's crust. Structures consists of a *geometric arrangement* — of planes, lines, surfaces, rock bodies, etc. The form and orientation of this arrangement reflect the interaction between the deforming forces and the pre-existing rock body.

Description and classification of structures. Because the geometry of structures is so important, a large body of descriptive nomenclature and classification has grown up, which it is essential to master if we wish to describe and understand structures. The arrangement of this book reflects my belief that there is little point in discussing such matters as stress, strain and processes of deformation before learning what it is that we wish to explain by such processes. Thus I have discussed the more descriptive aspects of structural geology (*morphology*) in Part 1 before proceeding to discuss deformation in Part 2.

Deformation. This is the process that changes the shape or form of a rock body—in other words, the process responsible for the formation of geological structures. To understand deformation, it is necessary to understand *stress* and *strain*, which deal with the manner in which material reacts to a set of forces. We must also discuss the *behaviour of materials*, since the way that a rock deforms is dependent on the physical properties of different rocks and on how these change with changes in temperature and pressure, and with time. We can then apply the principles of deformation to the formation of specific types of structures such as faults and folds.

Geotectonics. In the third section of the book, I have attempted in a very selective way to show how structures and deformation may be related to large-scale Earth processes. The subject of geotectonics essentially covers large-scale structural geology—that is, the study of large Earth structures such as mountain belts and continental margins. The advent of *plate tectonics* has meant that many types of hitherto unrelated geological phenomena can be explained in terms of a unifying theory of crustal movements and processes. It is essential for the structural geologist to see individual structures or deformed areas in their context and to try to relate them to some large-scale pattern, even if the attempt subsequently proves to have been a failure. Only in this way will our understanding of the Earth progress.

Other related topics. There are some topics which I have omitted, partly because I wished to keep the book as short as possible, and partly because they are only of marginal concern and are adequately dealt with in other textbooks. One of the more important of these is *sedimentary structures.* Structures produced as a result of processes associated with sedimentation are not of great concern to most structural geologists, who are more interested in the deformation of solid rocks. However, there are areas of overlap between sedimentary and structural geology that should not be forgotten. One important problem for the field geologist is distinguishing between sedimentary and deformational structures. This problem is particularly acute in highly deformed metamorphic terrains where the origin of early and poorly-preserved structures is often unclear. Fold-type structures produced by soft-sediment slumping and other processes are open to misinterpretation. When fold-type structures are confined to a single layer (particularly if they are truncated by the layer above) they are likely to be of sedimentary origin and must be treated with caution.

Many sedimentary structures are, of course, of indirect interest to the structural geologist since they reflect tectonic control. Thus features indicating slumping or sliding of soft or unconsolidated sediments are often a direct result of tectonic processes such as earthquakes, fault movements, etc. Certain sedimentary structures are also of value to the structural geologist as indicators of the younging direction of the strata. Cross-bedding, graded bedding and other 'way-up' structures have been used extensively in highly deformed terrains to elucidate the large-scale structure (see Figure 2.5).

Another important topic that is avoided is the interpretation of geological structures from maps and aerial photographs. I hope that students who read this book will have had, or will gain, some experience in this subject since it is essential to the three-dimensional appreciation of structural geometry. There are a number of excellent textbooks on geological maps and air photo interpretation that the reader may consult. However, certain basic stratigraphic and geometric concepts relating to map interpretation of structure are outlined below.

Stereographic projection, an essential geometric tool in structural geology, is dealt with in the Appendix.

Warning to students! It is easy for a student to be misled into regarding what is printed in a textbook as unquestioned and immutable truth. However, in geology, perhaps more than in other sciences, today's facts may become tomorrow's discarded theories. Much of the material of this book is based on the opinion of established experts based on sound evidence. Some of it is disputed. Be sceptical!

Many textbooks attribute all statements to their original author by liberal reference to published work. In this way an impression is conveyed of a changing body of fact built up by numerous individual scientists, sometimes with opposing ideas. This is of course the correct scientific procedure. However, I have chosen in this short book not to follow this procedure because I feel that large numbers of references break up the smooth flow of the text and make for less easy comprehension. Instead, I have listed selected references for further reading at the end of each section, hoping that the reader will be encouraged to sample some of the original contributions to the subject and to proceed to more advanced texts.

Basic concepts

Bedding or stratification: definitions and geometry. A *bed* is a layer of rock deposited at the Earth's surface and bounded above and below by distinct surfaces (*bedding planes*); these usually mark a break in the continuity of sedimentation, i.e. a cessation of sedimentation, or a period of erosion, or a change in type or source of sediment. Beds are normally sedimentary, but may also consist of volcanogenic material. A thickness in the range cm to m is normally implied. 'Bed' is more or less synonymous with *stratum*, but the latter term is normally used only in the plural (e.g. 'Silurian strata'). Beds may be relatively homogeneous in composition and internal structure, and represent more or less continuous deposition. However, the term is also used for a sedimentary unit composed of numerous thin distinct layers. The term *bedded* means composed of beds: thus 'bedded rocks', 'thin-bedded', 'cross-bedded' etc.; *bedding* is used as a collective noun for the beds in a particular outcrop or area: thus, 'the bedding dips to the west'. The term is also used to describe various characteristics of the beds, such as 'cross-bedding' and 'graded bedding'.

The simplest type of bedding geometry consists of a set of parallel planes, representing a group of beds, or a *formation*, of uniform thickness. However, in practice, beds and formations vary laterally in thickness, in which case the geometry of the formation must be described by two non-parallel bounding surfaces. Thickness variation in such a formation may be described by a set of *isopachytes* (see below).

Onlap and offlap. Onlap is the term used to describe a structure formed where successive wedge-shaped beds extend further than the margin of the underlying bed, such that they lie partly on older basement (see Figure BC.1A). Such a structure is typical of sedimentary sequences in expanding basins with transgressive shorelines. The term *overlap* is synonymous. *Offlap* is the structure formed where successive wedge-shaped

beds do not extend to the margin of the underlying bed, but terminate within it (see Figure BC.1B). Such a structure is typical of sedimentary sequences in contracting basins with regressing shorelines.

Figure BC.1 Cross-sections showing *onlap* and *offlap. A, onlap* of successive beds 1–4, each resting partly on older basement; *B, offlap* of successive beds 1–4, associated with a regressive shoreline.

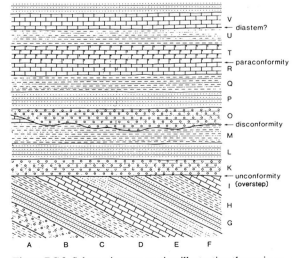

Figure BC.2 Schematic cross-section illustrating the various types of *stratigraphic break*. From Roberts (1982).

Unconformities and allied structures. Breaks in the stratigraphic record, representing intervals of geological time not marked by sediment deposition, are known variously as *diastems, non-sequences, paraconformities, disconformities* and *unconformities* (see Figure BC.2). *Diastems* represent pauses in sedimentation, marked by abrupt changes in sediment type, producing surfaces of discontinuity (bedding planes) but no other evidence of a time gap. *Non-sequences* (or *paraconformities*) are similar to diastems but exhibit faunal or other evidence of a time gap. *Disconformities* are marked by evidence of erosion during the sedimentary break, but the bedding below the erosion surface is parallel to that above, i.e. there has been no deformation of the lower series of beds prior to erosion.

Unconformities are distinguished from other stratigraphic breaks by angular discordance between the older beds below the unconformity surface, and the younger beds above. Hence an unconformity represents the following sequence of events:

(i) deposition of lower strata
(ii) tilting or other deformation of lower strata
(iii) erosion
(iv) deposition of upper strata.

The structure produced by the discordance of younger upon older strata is termed *overstep*, and the basal beds of the younger series are said to 'overstep' the various strata of the older series truncated by the erosion surface (see Figure BC.2). A *non-conformity* is a type of unconformity where younger strata rest on an erosion surface cut across non-bedded igneous rocks.

Geometry of inclined planes and lines. The attitude of an *inclined plane* such as bedding, foliation, faults etc. is conventionally described in terms of the *strike* and *dip* of the plane (see Figure BC.3A). The *strike* is the unique direction of a horizontal straight line on the inclined plane and is recorded as a bearing (azimuth). The *dip* is the inclination or tilt of a planar surface, e.g., bedding or foliation. The *true dip* of a plane (the angle from the horizontal to the plane) is measured in a vertical plane perpendicular to the strike, and is the maximum angle from the horizontal that can be measured for a given plane. Lines in any other orientation on the plane are at a smaller inclination to the horizontal; these angles represent apparent dips. The *apparent dip* is thus the angle of inclination of a given plane with the horizontal, measured in a plane that is not orthogonal to the strike. The angle of apparent dip measured in a series of vertical planes varies from zero (parallel to the strike) to a maximum in the direction of true dip. If the angle of apparent dip in two different directions is measured, the true

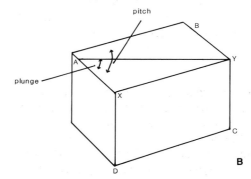

Figure BC.3 Inclined planes and lines. *A, strike* and *dip* of an inclined plane: the *true dip* of the plane ABCD is the angle XAD: the angle YAC is an *apparent dip*; B, *plunge* and *pitch* of an inclined line: the *plunge* of the line AC is the angle YAC: the *pitch* is the angle BAC.

dip can be calculated using a stereogram (see Appendix).

The *direction of dip* (i.e. the direction in which the plane dips downwards from the surface) is measured either directly as a compass bearing (azimuth) or in relation to the strike direction, which is 90° from the dip direction. Thus a bed may be said to dip at 30° SE, if the strike direction is specified, or at 30° to 110° if it is not. A dip arrow is the conventional representation of dip direction on a geological map. The position of the observation is indicated by the arrowhead, and the amount of dip in degrees is placed alongside the arrow. On many maps, this symbol is replaced by a line parallel to the strike, with a short tick indicating the dip direction.

The orientation of a *linear structure* (e.g. a fold axis) is measured in terms of *plunge* or *pitch*. The *plunge* is the angle between the line and the horizontal in the vertical plane. The plunge is given as an angle and a bearing (azimuth), which is the direction of plunge, thus, 30° to 045° or 30° NE. The *pitch* is the orientation of a line, measured as an angle from the horizontal, in a specified non-vertical plane. A measurement of pitch must give the strike and dip of the plane of measurement, plus the angle of pitch and the strike direction from which the pitch angle is measured (since there are two possible directions in a given plane for the same pitch angle)—see Figure BC.3B. This method is useful in the field where precise measurements of angles within inclined joint, foliation or bedding planes are more convenient than direct measurement of the plunge. The plunge may be easily derived using a stereogram (see Appendix). The instrument used in the field to measure the inclination (dip) of a planar surface or the plunge of a lineation is termed a *clinometer* and is often combined with a compass in order to measure the orientation of planes or lines with reference to geographic co-ordinates.

Representation of structures on geological maps. On geological maps, the attitude of planar beds etc., may be recorded by a set of *strike lines* drawn parallel to the strike of the plane or set of planes in question. If the dip of the planes is constant, the

Figure BC.4 *Strike lines* and *structure* (*stratum*) *contours. A, strike lines* at heights of 0 to 500 m on the inclined plane (coloured) project as a set of parallel *strike lines* (labelled 0–0, 1–1 etc.) on the horizontal plane (map); the spacing is closer on the steeper part than on the shallower part of the inclined plane; *B,* an inclined cylinder (plunging fold) intersects a set of horizontal planes at heights of 0–500 m in curved lines termed *structure* or *stratum contours*: these project on the horizontal plane (map) as shown to give a map representation of the shape of the structure.

strike lines are straight, with uniform spacing. An increase in dip produces a decrease in spacing and a decrease in dip produces an increase in spacing (see Figure BC.4A). A surface with variable strike is represented by curved strike lines known as *structure contours* (Figure BC.4B). Structure contours follow a constant height on a geological surface, and a set of such contours, drawn at uniform height intervals, represents the three-dimensional shape of the surface in the same way that topographic contours represent the height variation of the land surface on a topographic map. A less precise but often more convenient method of portraying the structure on a map is to use *form lines*. These are lines drawn on a map to indicate the general direction of the strike of a folded surface (e.g. see Figure 2.13). A tick on the line indicates the dip. A set of form lines will illustrate the geometry of the folding in a similar way to the outcrop pattern of the strata, for example, but are more precise, and are not affected by topography. They can therefore be used in areas where individual formations have not been mapped. A set of form line contours can be drawn in a precise manner such that the spacing is proportional to the dip. A contoured map constructed by this means will illustrate the shape of a folded surface in the same way that a topographic contour map displays topographic relief (see Ragan, 1973). A *form surface* is any planar surface that intersects the ground surface as form lines and which may be used for structural mapping.

An *isopachyte* is a line joining points of equal stratigraphic thickness of a formation or group of strata. An isopachyte map is contoured to indicate the three-dimensional shape of a unit of variable thickness. The technique is used for example in the study of sedimentary basins and in portraying the geometry of stratigraphic units cut off by unconformities. Isopachyte maps may be prepared from borehole data from which thicknesses are directly obtainable, or by geometric construction using stratum contours for the top and base of the unit, and subtracting the lower from the higher values where they intersect. The line of intersection of a stratigraphic boundary with a higher stratigraphic boundary such as an unconformity is marked by the zero isopachyte of the rock body between the two boundaries in question. This line is often termed the *feather edge*, e.g. 'the feather edge of the base of the Coal Measures on the base of the Triassic'.

A related term is *subcrop*, which is the subsurface 'outcrop' of a rock unit. A stratigraphic formation may intersect a sub-surface plane, e.g. an unconformity or fault, in a subcrop, which represents the area of the plane lying between the lines of intersection (*feather edges*) of the boundaries of the formation.

Topographic effects. In areas of horizontal or gently-dipping strata, outcrop patterns are controlled mainly by the topographic relief. Younger beds occur at higher topographic levels and older at lower levels. Outcrops of younger rock completely surrounded by older are termed *outliers*, and correspond to hills separated by erosion from other outcrops of the same beds (Figure BC.5). Conversely, an *inlier* is an area of older rocks surrounded by younger, e.g. in a valley cut through younger strata. Both outliers and inliers may be created by the interaction of structures with topography. For example an anticline or a horst (see Figure 1.9) crossing a valley may form respectively a fold or a fault inlier.

It is usually necessary to supplement the two-dimensional information on the geological structure of an area provided by the geological map by *cross-sections*, which are diagrammatic representations (normally constructed in the vertical plane) of the geology of an area. Reasonable assumptions are made about the way in which structures visible at the surface continue down-

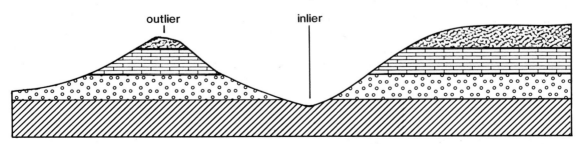

Figure BC.5 *Outlier* and *inlier* produced by the intersection of horizontal strata with topographic relief: the hill produces an outcrop of younger rock surrounded by older (*outlier*) and the valley an outcrop of older surrounded by younger (*inlier*).

wards, and the surface information may be supplemented by data from boreholes, wells, etc. Cross-sections may be drawn along a particular line or lines on the map, chosen to illustrate the vertical structure most effectively. The combination of map and cross-section should ideally give a good three-dimensional picture of the geological structure of an area. In complex areas, several lines of section may be used to give a better coverage of the structural variation.

Instead of a vertical section, a *down-plunge projection* may be employed; this is a reconstructed profile or cross-section of a fold structure drawn perpendicular to the plunge of the fold axis. This is done to give a more accurate representation of the fold geometry (see section 10.2).

It is important in interpreting the history of an area to be able to visualize the original geometry of a set of rocks before deformation. A geometrical reconstruction, in the form of a map or cross-section, is often employed for this purpose, and is termed a *palinspastic* reconstruction. *Balanced sections* (see section 8.7) are a special type of palinspastic reconstruction much used in the interpretation of complex fold/fault belts.

Further reading

Butler, B.C.M. & Bell, J.D. (1988) *Interpretation of Geological Maps*, Longman Scientific and Technical, Harlow, Essex.

Phillips, F.C. (1971) *The use of Stereographic Projection in Structural Geology*, Edward Arnold, London.

Roberts, J.L. (1982) *Introduction to Geological Maps and Structures*, Pergamon, Oxford.

Ragan, D.M. (1973) *Structural Geology: an Introduction to Geometrical Techniques (2nd edition)*, Wiley, New York.

Part 1: Morphology— how to describe and classify structures

In this part of the book, the various kinds of geological structure are described and classified in terms of their shape or form (*morphology*). We shall deal in turn with *faults and fractures, folds, foliation, lineation and fabric* and *igneous bodies*.

1 Faults and fractures

1.1 Rock fractures

Rock fractures are the commonest type of geological structure, and may be seen in any rock exposure. They are cracks across which the cohesion of the material is lost, and may be regarded as planes or surfaces of discontinuity. Where there is a measurable displacement across the fracture plane, that is, where the rock on one side has moved along the fracture relative to the other side, the fracture is termed a *fault*. Where there is no displacement, or where the displacement is too small to be easily visible, the fracture is termed a *joint*. Although the distinction between the two appears somewhat artificial, and depends on the scale of observation, in practice, the great majority of fractures show no, or negligible, displacement and are classified as joints.

Fractures are important in a number of ways. Their presence significantly affects the strength of a rock, and they must be carefully studied in civil engineering operations such as those involved in the construction of tunnels and dams. They are also important sites of mineralization, since dilational fractures developed under *extensional stress* are normally occupied by vein material deposited in the space created as the fracture opens.

1.2 Fault geometry and nomenclature

Definition of a fault. A fault is a planar fracture across which the rock has been displaced in a direction that is generally parallel to the fracture plane.

Geometry of displacement. The main elements of the displacement geometry of a fault are shown in Figure 1.1. Where the fault plane is non-vertical, the block above the fault is referred to as the *hangingwall* block and the block below the fault as the *footwall* block. The inclination of a fault plane may be given as a *dip*, in the same way as bedding (see Figure BC.3A), but is often measured as the angle between the fault plane and the vertical, in which case it is termed the *hade*. The displacement of the fault plane between the two blocks may take any direction within the fault plane. Faults with a displacement parallel to the strike of the fault plane are termed *strike-slip* faults and those with a displacement parallel to the dip of the fault plane are termed *dip-slip* faults. Faults with oblique-slip displacements are regarded as having strike-slip and dip-slip components, as shown in Figure 1.1. Strike-slip faults may also be called *wrench, tear* or *transcurrent* faults.

The measurement of the displacement on dip-slip faults is often made with reference to the horizontal and vertical components of the displacement, which are termed respectively the *heave* and the *throw* (Figure 1.2). It is the throw,

Figure 1.1 Components of fault displacement: *ss*, strike-slip component; *ds*, dip-slip component; *PP'*, true displacement vector.

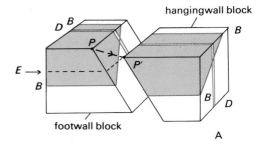

Figure 1.2 Geometry of dip-slip fault displacement: h, heave; t, throw; α, angle of dip; θ, angle of hade $(= 90° - \text{dip angle})$; ds, dip-slip component of displacement.

or vertical displacement, that is normally quoted for a dip-slip fault rather than the true displacement. The relationship between these elements is given by

$$\tan \alpha = \frac{\text{throw}}{\text{heave}} = \frac{t}{h}$$

and

$$\sin \alpha = \frac{\text{throw}}{\text{true displacement}}$$

$$= \frac{t}{ds} \tag{1.1}$$

where α is the dip of the fault.

It is important to realize that fault displacements are difficult to measure in practice because it is frequently impossible to match precise points on each side of the fault. If bedding is displaced, we cannot be certain how much of the apparent displacement is due to dip-slip and how much to strike-slip movements (Figure 1.3*A*, *B*). The problem is overcome if the direction of movement on the fault plane is indicated by *slickenside striations* (see section 1.4) or if there is a measurable offset on a vertical structure, such as a dyke, which can be used to measure the strike-slip component (Figure 1.3*C*).

Sense of displacement. The sense of relative displacement on faults is important and depends upon the orientation of the fault with respect to the stress axes (see section 9.2). In the case of strike-slip faults, the displacement is termed *sinistral* (or left-lateral) if the opposite block moves to the left, and *dextral* (or right-lateral) if the opposite block moves to the right, as viewed by an observer standing on one side of the fault (Figure 1.4).

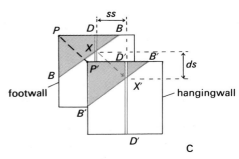

Figure 1.3 Measurement of fault displacement. (*A*) The fault affects dipping bedding *B–B* and vertical dyke *D–D*. The true displacement vector is *P–P'*. (*B*) Map at erosion level *E* of *A* shows horizontal displacement of bedding and dyke along fault. Note that the amount of displacement appears to be different. The true horizontal displacement is shown by the vertical dyke *D–D*. (*C*) View of fault plane looking down from the right, showing the trace of bedding *B–B* and dyke *D–D* on the footwall displaced to positions *B'–B'* and *D'–D'* on the hangingwall. The true displacement *P–P'* is given by the movement of intersection *X* of *B–B* and *D–D* to position *X'*. The strike-slip component *ss* is given by the displacement of the vertical dyke *D–D* to *D'–D'*. The dip-slip component *ds* must be measured using both dyke and bedding displacements.

In the case of dip-slip faults, the displacement is termed *normal* when the hangingwall moves down and *reverse* when the hangingwall moves up, relative to the footwall (Figure 1.5). An alternative way of expressing the displacement

Figure 1.4 Sinistral and dextral displacements on strike-slip faults (plan view).

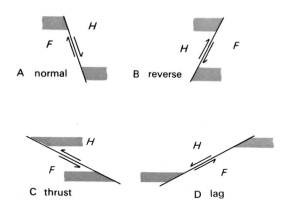

Figure 1.5 Normal and reverse displacements on dip-slip faults (vertical sections). *F*, footwall; *H*, hangingwall. *A*, normal fault; *B*, reverse fault; *C*, thrust; *D*, lag (low-angle normal fault).

in dip-slip faults is to refer to the direction of throw. The direction of dip of normal faults is towards the *downthrown* side, whereas in reverse faults the dip is directed towards the *upthrown* side. The sense of displacement in a reverse fault results in lower rocks (normally older) being placed above higher (normally younger), whereas the opposite is true in normal faults.

Thrusts and lags. Faults dipping at less than 45°, or low-angle faults, are distinguished from high-angle, dip-slip faults. If the sense of movement is reverse they are termed *thrusts*, and if the sense of movement is normal they are termed *low-angle normal faults*, or *lags*. Thrusts are particularly important in orogenic belts and often have displacements of many tens of kilometres. The Moine thrust which marks the north-western margin of the Caledonian orogenic belt in north-western Scotland (see Figure 15.7) has an estimated displacement of about 100 km. *Thrust, extensional* and *strike-slip fault systems* involve complex geometries and nomenclature, described in sections 9.4–9.6.

1.3 Rocks produced by faulting (fault rocks)

Fault breccia and gouge. Many faults are marked by a zone of broken and crushed rock fragments of varying size. This material is called *fault breccia* where the visible fragments make up an appreciable proportion of the rock. Where the bulk of the rock consists of fine powder, the material is termed *fault gouge.* Since such zones are normally softer and more easily eroded than the unfaulted rock, they give rise to the marked topographic depressions often associated with fault outcrops.

Cohesive crush rocks. Fault breccia and gouge are essentially loose-textured fault rocks found near the surface. At greater depth, various kinds of cohesive crush rocks are found where the rock is lithified and the increased pressure has, in many cases, caused partial recrystallization of the rock texture. Such rocks are termed *crush breccias*, where visible fragments dominate the rocks, or *cataclasites* where the fine-grained matrix makes up an appreciable proportion of the rock.

Mylonites. Structural geologists generally attempt to distinguish rocks formed under brittle conditions by breaking and crushing of the material (*cataclasis*) from those formed under ductile conditions by continuous recrystallization or flow (see section 7.3). Finer-grained rocks produced by the latter process are hard and 'flinty' with a platey or streaky texture. Such rocks are termed *mylonites* (Figure 1.6A). Where recrystallization is dominant, the rock is termed *blasto-mylonite*.

Ultramylonite and pseudotachylite. Extreme crushing produces a rock composed of broken fragments in a dark, often black, matrix of ultramicroscopic grains. Such material is termed *ultramylonite* (Figure 1.6A). Frictional heating caused by rapid relative movement along the fault may be sufficient to melt some of this material, forming a glassy substance, often containing spherulites, termed *pseudotachylite*, which forms veins intruding the adjacent frac-

Table 1.1 Classification of fault rocks. From Sibson (1977).

		Random-fabric	Foliated	Proportion of matrix
Incohesive		Fault breccia (visible fragments > 30% of rock mass)	?	
		Fault gouge (visible fragments < 30% or rock mass)	?	
Cohesive	Glass/devitrified glass	Pseudotachylite	?	
	Nature of matrix — Tectonic reduction in grain size dominates grain growth by recrystallization and neomineralization	Crush breccia (fragments > 0.5 cm) / Fine crush breccia (0.1 cm < frags. < 0.5 cm) / Crush microbreccia (fragments < 0.1 cm)		90–100%
		Protocataclasite	Protomylonite	50–90%
		Cataclasite	Mylonite	10–50%
		Ultracataclasite	Ultramylonite	0–10%
	Grain growth pronounced	?	Blastomylonite	

(Cataclasite Series — Phyllonite varieties — Mylonite series)

tured rock (Figure 1.6*B*, *C*). Since the glassy material is usually devitrified, and contains a high proportion of unmelted fragments, it is often difficult to distinguish from ultramylonite except under high magnification. Pseudotachylite is apparently formed only at depth in the crust under moderate load pressure and a relatively rapid deformation rate. Thus a fault exhibiting soft gouge at the surface might develop pseudo-tachylite at intermediate depths, and at deeper levels might be replaced by a mylonite.

1.4 Features associated with fault planes

Slickensides. Fault planes frequently show shiny or striated surfaces caused by the rubbing or polishing action of the opposite face as it moved across. Such features are termed *slickensides*, and

Figure 1.6 Crush-rocks produced by faulting. *A.* Gradational change from unaltered rock on the right through mylonite to ultramylonite on the left in amphibolite from Ness, Lewis. Plane polars, × 14 (from Sibson, 1977, plate 3). *B.* Discordant intrusive veins of dark pseudotachylite cutting foliated gneiss (from Gairloch, NW Scotland). Plane polars, × 5. *C.* Devitrified spherulitic structure in pseudotachylite from vein cutting Lewisian gneiss at Gairloch, NW Scotland. Plane polars, × 250. *B* and *C* from Park, R.G. (1961) *Am. J. Sci.* **259**, 542–550, plate 1.

Figure 1.7 Hand specimen showing slickenside striations on a fault surface. Scale bar, 6 cm.

the grooves or striations indicate the direction of relative movement of the fault (Figure 1.7). Other linear structures are due to the growth of fibrous minerals such as quartz or calcite that have their long axes parallel to the direction of motion.

Flexures associated with faults. Layered rocks adjoining a fault commonly exhibit open folds or flexures which appear to be related to the fault movement. Such folds have been termed

'*drag folds*' by some authors. This term should be used with caution since it suggests that the fold originated by the dragging action of the opposing fault block, whereas many flexures along fault planes result from an initial ductile strain that precedes fracturing (Figure 1.8).

A before fracture   B after fracture

Figure 1.8 Association of faults and flexures. Ductile bending in a strained zone may precede the fault movement causing flexures in the rocks adjacent to the fault.

Figure 1.9 Graben and horst structure.

Figure 1.11 Transference of displacement from thrust to strike-slip fault.

1.5 Fault associations

Faults are usually found in groups of the same type. Thus a major fault with a large displacement may be accompanied by a set of smaller parallel faults with the same sense of displacement. Major uplifted or depressed blocks may be bounded by sets of faults of the same type but opposite sense of movement. A depressed block bounded by normal faults is termed a *graben* and an elevated block a *horst* (Figure 1.9). Major graben features which extend for long distances are called *rifts*. Two well-known examples are the great African rift and the Rhine rift, both of which consist of a series of connected graben extending for hundreds of kilometres.

Fault associations need not consist of parallel faults. Quite complex arrangements of branching faults may accompany a major fault (see Figure 15.3) and very often the termination of a fault is marked by branching *splay faults* which

Figure 1.10 Splay faults at the termination of a main fault. The displacement on the main fault may be spread over a large area around the termination of the main fault by the cumulative effect of displacements on smaller branching faults.

spread the displacement over a large area (Figure 1.10).

Faults may also terminate against other faults of different type; for example, two thrusts may be linked by a strike-slip fault so that the displacement is transferred from one to the other. Such a fault is known as a *transfer fault* (Figure 1.11).

1.6 Joints

Joints may occur in sets of parallel, regularly-spaced fractures and several sets may occur in the same rock, giving a conspicuous blocky appearance to an outcrop (Figure 1.12). More commonly, however, joints are much less regular and systematic. Where a recognizable joint set exists, it can usually be related in some way to the tectonic stresses and to the geometry of the rock body containing the joints. For example, joint sets are frequently found both perpendicular and parallel to the bedding in layered rocks. The perpendicular joints may form two or more intersecting sets which bear a simple relationship to the regional folds (Figure 1.13).

'Unloading' joints. Many joints are due to the release of 'stored' stress. The weight of a great thickness of overlying strata causes deeply buried rock to be compressed. However, once the overlying rock has been eroded, this load pressure is reduced and the rock expands by the development of tensional joints which are often parallel to bedding surfaces in sedimentary strata, or to

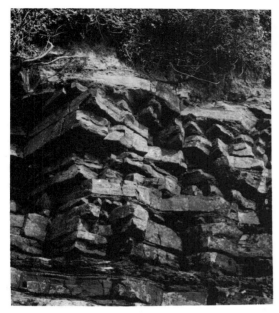

Figure 1.12 Joint sets in thin-bedded Devonian sandstones, Lligwy bay, Anglesey.

Figure 1.14 Polygonal cooling joints forming columnar structure in basalt, Giant's Causeway, Co. Antrim. From Holmes, A., *Principles of Physical Geology*, Nelson, 1978, Fig. 5.7 (photograph J. Allan Cash).

the contemporary erosion surface in massive igneous rocks, where they are termed *sheet joints*.

Cooling joints. Another common cause of joint formation is the contraction which takes place in a cooling igneous body. Tabular igneous bodies, i.e. dykes and sills, frequently exhibit polygonal columnar jointing perpendicular to the cooling surfaces. The spectacular columnar jointing of the Giant's Causeway in Antrim (Figure 1.14) is a well-known example of this type of structure.

Relationship between joints and regional deformation. Under favourable circumstances, regu-

lar joint sets that occur regionally in various different rock types can be related to a regional compression or extension in the same way as folds. The way in which fractures are related geometrically to stress is discussed in detail in section 9.2.

Further reading

Hobbs, B.E., Means, W.D. and Williams, P.F. (1986) *An Outline of Structural Geology*, 2nd edn, John Wiley, New York. An excellent general textbook.

Sibson, R.H. (1977) Fault rocks and fault mechanisms. *J. geol. Soc. Lond.* **133**, 191–213. Gives a useful classification of fault rocks and discusses the relationship between mylonite, pseudotachylite and other fault rocks.

Suppe, J. (1985) *Principles of Structural Geology*. Prentice-Hall, Englewood Cliffs, New Jersey. Chapters 6 and 8 give considerable additional information about faults and particularly joints, with numerous illustrations.

See also section 9 references.

Figure 1.13 Relationship of joint sets to major folds. Regular joint sets perpendicular to bedding may be divided into longitudinal (parallel to fold axes), transverse (perpendicular to fold axes), and diagonal (oblique to fold axes).

2 Folds

While fractures are the commonest expression of rock deformation, there can be no doubt that folds are the most spectacular. When we see a bed of rock, which was originally flat and planar, now bent into a huge arch many hundreds of metres in height, we are forced to recognize in a very direct way the existence of the tremendous forces that act upon the Earth's crust. The sight of a large-scale fold in a mountainside, as can be seen in the Alps and other mountain belts, is much more satisfying as an expression of deformation than reconstructions based on mapping, although the latter may be equally convincing.

2.1 Meaning and significance of folds

A *fold* is a structure produced when an originally planar surface becomes bent or curved as a result of deformation. We have seen that fractures result from brittle deformation that causes the rock to break completely along discrete planes. Folds, however, are an expression of a more ductile type of deformation which produces gradual and more continuous changes in a rock layer, both in its attitude and internally, as the rock accommodates to changes in shape. Such a deformation is more pervasive than that res-ponsible for faulting. The difference between brittle and ductile deformation and the way rocks accommodate physically to deformation are dis-cussed in section 7.

2.2 Basic fold geometry and nomenclature

Fold hinge and limbs. If we consider a single folded surface (Figure 2.1*A, B*), the main ele-ments of the geometry of the fold shape are, firstly, the *hinge* or *closure*, which is the zone of

maximum curvature of the surface, and, second-ly, the *limbs*, which are the areas between the hinges. If it is possible to define a line of maximum curvature, this line is called the *hinge line*. A single fold comprises a hinge and two limbs which enclose the hinge. In a series of folds, each limb is common to two adjacent folds. Depending on the actual shape of the fold, that is, on the way in which a surface changes curvature, the hinge may be very sharply defined and the limbs relatively straight (Figure 2.1*B*), or the curvature may be more constant around the fold if it approaches a cylindrical shape (Figure 2.1*C*). In a truly *cylindrical fold*, where the fold surface corresponds to part of the surface of a cylinder, neither hinge nor limb can be defined, but such examples are uncommon. The term 'cylindrical' used in this sense should not be confused with '*cylindroidal*' which is defined in section 2.6.

Fold axis and axial plane. In describing the attitude of a fold, it is useful to refer, not to the attitudes of both limbs, but to the imaginary plane that is equidistant from each limb and which bisects the angle between them. This plane is termed the *axial plane* and cuts the hinge zone of the fold along a line termed the *fold axis* (Figure 2.2*A*). Provided that the limbs of a fold are sufficiently well defined, the determination of the orientation of axial plane and axis will give a precise description of the orientation of the fold. It must be remembered that these terms refer only to a single fold surface.

Fold axial surface. When we consider a fold consisting of a number of fold surfaces, it is usually more convenient to define a single surface

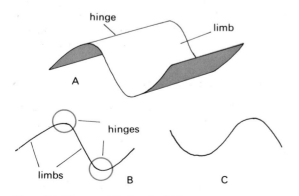

Figure 2.1 Hinge and limbs of a fold.

Figure 2.2 Axis, axial plane and axial surface of a fold.

such that it passes through the hinge lines of each successive fold surface. Such a surface is in general not planar and is known as the *fold axial surface* (Figure 2.2*B*). The fold axial surface does not, in general, correspond with the axial planes of the individual layers in the fold. The hinge lines are not always easy to locate by eye if the hinge zone is broad, and the method is only appropriate for relatively tight folds with well-defined hinges.

Inter-limb angle. The smaller angle made by the limbs of a fold is termed the *inter-limb angle* or *fold angle*. The tightness or openness of a fold as expressed by this angle is a useful method of classifying folds (Figure 2.7*A*) and reflects the amount of deformation.

Amplitude and wavelength. These are convenient measures of the *size* of a fold (Figure 2.3). The easiest method of determining the *wavelength* of a fold is to measure the distance between the hinges on each side of the fold. If these are not visible, the half-wavelength may be measured instead by taking the distance between the two *inflexion points* on each side of the fold hinge. The inflexion points are where the sense of curvature changes from one fold into the next (see Figure 2.7*B*). The *amplitude* or 'height' of a fold may be measured by taking half the perpendicular distance from the hinge to the line between the two adjoining hinges or the perpendicular distance from the hinge to the line between the two inflexion points.

Figure 2.4 Plunge of a fold. The angle of plunge is measured in a vertical plane from the horizontal.

2.3 Fold orientation

The orientation of a fold may be measured in the field by obtaining the attitudes of both limbs, and if the limbs are relatively straight and regular this information is sufficient for a complete description of the fold attitude. However, it is more convenient, and conveys more useful information, to describe and record folds in terms of their *axial planes* or *surfaces*, and *axes*. Thus on a map, the position of a fold is shown by drawing the outcrop of the fold axial surface as a line. This line is termed the *fold axial trace*. The dip and strike of the axial plane or axial surface may also be plotted in the same way as for a bedding plane.

Fold plunge. The attitude of the fold axis is measured as the angle between the axis and the horizontal. This angle, which must be measured in a vertical plane (like the dip angle) is termed the *plunge* (Figure 2.4), so that a complete description of the orientation of a fold axis is given as an amount and direction of plunge; for example, a fold plunges 30° towards (on a bearing of) 105°.

2.4 Classification of folds

The classification of folds is based on four main features or properties: direction of closing, attitude of axial surface, size of inter-limb angle and nature of profile.

Closing and facing direction. Folds that close upwards, that is where the limbs dip away from

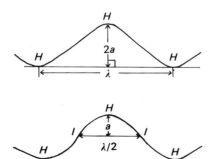

Figure 2.3 Amplitude and wavelength of a fold. λ, wavelength; *a*, amplitude; *H*, hinge; *I*, inflexion point (see text).

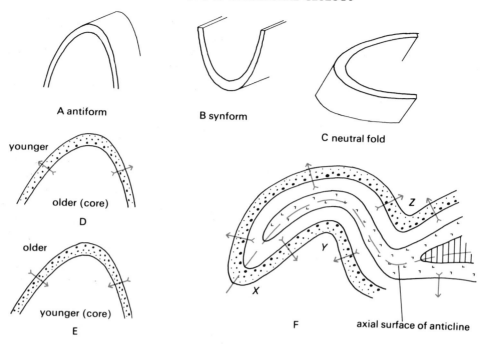

Figure 2.5 Closing and facing directions of a fold. *A*, antiform—an upward-closing fold; *B*, synform—a downward-closing fold; *C*, neutral fold—a sideways-closing fold; *D*, anticline—older rocks in core. Fold faces upwards. *E*, syncline—younger rocks in core. This fold faces downwards and is an antiformal syncline. *F*, refolded recumbent anticline—*X* is a synformal anticline, *Y* is an antiformal syncline, and *Z* is a synformal syncline.

the hinge, are termed *antiforms* (Figure 2.5*A*) and those that close downwards, where the limbs dip towards the hinge, are termed *synforms* (Figure 2.5*B*). Folds that close sideways are termed *neutral folds* (Figure 2.5*C*). Under normal conditions where the bedding becomes younger upwards, an antiform will contain older rocks in its core and is then given the more familiar name *anticline*. Thus the term anticline should strictly be applied only to a fold with older rocks in its core (Figure 2.5*D*). Conversely, a *syncline* is a fold that contains younger rocks in its core (Figure 2.5*E*).

In areas of more complex folding, where strata are commonly inverted, it is possible to find downward-closing anticlines or upward-closing synclines (Figure 2.5*F*). In such areas it is convenient to speak of the *facing* direction of a fold, which is the direction along the axial surface in which the strata become younger. Thus in Figure 2.5*F*, fold *X* is a synformal anticline and

fold *Y* is an antiformal syncline, and both folds face downwards since the strata are inverted. The main fold, with a sub-horizontal axial surface, is a neutral fold, facing towards the left. Fold *Z* is an upward-facing synform, i.e. a synformal syncline.

Attitude of axial plane. Folds may be divided into three groups based on the dip of the axial plane or surface. Folds with steep to vertical axial planes are termed *upright*, those with moderately-dipping axial planes are termed *inclined* and those with sub-horizontal axial planes are termed *recumbent* (Figure 2.6*A*–*D*). The division between these classes is not rigidly defined. Inclined folds where one limb is inverted are often termed *overfolds* (Figure 2.6*C*). In inclined folds, the highest and lowest points on the fold surface do not in general correspond with the hinges and it is sometimes useful to use the terms *crest* and *trough* respectively for these positions

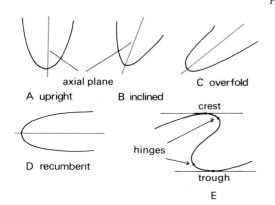

Figure 2.6 Attitude of the fold axial plane. *A*, upright; *B*, inclined; *C*, overfold; *D*, recumbent. *E*, distinction between crest and trough lines and hinge lines in an inclined fold.

(Figure 2.6*E*). The crest is of interest in petroleum exploration since it will form a trap for oil or gas in a permeable layer sealed by an impermeable layer above.

Inter-limb angle. The size of the inter-limb angle measures the degree of tightness of a fold and reflects the amount of compression of the strata. Figure 2.7*A* shows a classification scheme that sub-divides folds into four classes: *gentle* (180°– 120°), *open* (120°–70°), *close* (70–30°), *tight* (30°–0°) and *isoclinal* (0°). These limits are

intended only as a general guide; precision is better achieved, if required, by stating the angle. Gentle folds are often termed *flexures*. Problems arise in measuring the fold angle where the limbs are not straight. In such a case, tangents can be drawn to the curve at the point of inflexion (Figure 2.7*B*). The inter-limb angle is the angle made by the two tangents.

Profile. The fold profile is the shape of a folded layer observed in the plane perpendicular to the fold axis. Folds exhibit considerable variation in profile, and since this variation partly reflects differences in the mechanism of formation (see section 10), a precise description of this profile is often very important.

The main categories of profile are shown in Figure 2.8. The simplest is the parallel fold where the adjacent fold surfaces bounding the folded layer are parallel, that is, the thickness of the layer measured perpendicular to the fold surface (the *orthogonal thickness*) is constant (Figure 2.8*A*). A special case of parallel fold is the *concentric fold* where adjacent fold surfaces are arcs of a circle with a common centre, known as the *centre of curvature* of the fold (Figure 2.8*C*).

Parallel folds die out upwards and downwards as a consequence of their geometry. This pro-

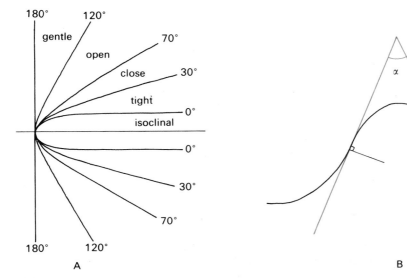

Figure 2.7 *A*. Classification of folds based on the inter-limb angle (after Fleuty, 1964). *B*. Measurement of inter-limb angle α in folds with rounded profile. Tangents are drawn to the fold surface at the points of inflexion.

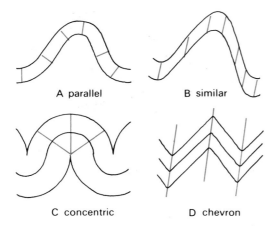

Figure 2.8 Types of fold profile. *A*. Parallel—coloured lines show constant layer thickness measured perpendicular to fold surface. *B*. Similar—coloured lines show constant layer thickness measured parallel to axial surface. *C*. Concentric—coloured lines are radii of a circle. *D*. Chevron—coloured lines are kink planes (axial surface traces) separating straight fold limbs.

perty is especially obvious in the case of concentric folds (Figure 2.8*C*) where the extent of the fold is limited by the centres of curvature. Beyond the centres of curvature, the shortening would have to be accommodated by faulting or by a different type of folding.

Another type of fold is the *similar fold* where the orthogonal thickness of the folded layer changes in a systematic way such that the fold maintains a constant thickness measured parallel to the axial surface (Figure 2.8*B*). In a true similar fold, the shape of adjacent curves should correspond precisely, and this property enables a similar fold to maintain its shape indefinitely upwards or downwards through successive layers.

Folds which possess planar limbs and sharp angular hinges are known as *chevron* or *accordion folds* (Figure 2.8*D*). Such folds exhibit the curious property of being apparently both similar and parallel—in the sense that many of the individual folded layers may be parallel whereas the fold as a whole is usually similar. Where such a fold is markedly asymmetric, the superimposed short limbs give the appearance of bands running across the rock. Such bands, which are effectively contained within adjacent

axial surfaces, are known as *kink bands* (see Figure 2.12*B*).

These fold types represent end members of continuous series of natural folds. Thus concentric to chevron represents one series, from extreme rounded to extreme angular, and parallel to similar represents another. Folds may be described in terms of how closely they approach these ideal types. For more exact work in folds, it is necessary to record the changes in curvature in more detail, and methods for doing this are described in section 10.2.

2.5 Description of fold systems

The way in which folds are arranged in systems, and how they are related to each other, is just as important in understanding their process of formation as the shape of the individual folds.

Fold symmetry. A set of folds is regarded as *symmetric* if the limbs are of equal length, and *asymmetric* if the limbs are of unequal length (Figure 2.9). Asymmetry may also be expressed as differences in thickness between two sets of limbs. A *monocline* is a special type of asymmetric fold where one limb is very short in relation to the size of the other limbs. The term monocline is normally used only for large-scale folds produced by a local steepening in dip.

Parasitic folds. Very often a set of folds with a small wavelength is superimposed on folds of larger wavelength. The smaller folds, which occur on the limbs or hinge of the larger, are known as *parasitic* or *satellite folds* (Figure 2.10*A*). In many cases, there is a systematic relationship between the symmetry of the parasitic folds

Figure 2.9 Symmetric and asymmetric folds.

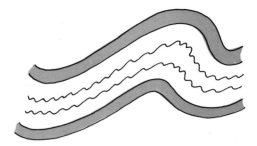

Figure 2.11 Disharmonic folds. The wavelength of the inner thinner layers is much shorter than that of the outer layers.

Figure 2.10 Parasitic folds and the use of the enveloping surface. *A.* Parasitic folds have an asymmetric 'Z' profile on the left-hand limb of the anticline, an 'M' profile on the crest and an 'S' profile on the right-hand limb. *B.* A complex profile may be simplified by drawing one or more enveloping surfaces joining the parasitic fold hinges.

and their position on the larger folds. Thus when both small-scale and large-scale folds are generated together, the sense of asymmetry will generally change from Z-shaped on the left-hand limbs of anticlines through M-shaped over the hinge area, to S-shaped on the right-hand limbs when viewed in profile. The existence of folds of different orders of magnitude can produce a very complicated shape in the folded surface. Locally measured bedding dips on such a surface will give no idea, or very little, of the overall shape of the main structure. However, this can be considerably simplified by the use of the *enveloping surface*, which is a surface drawn through the hinge lines of all the folds (Figure 2.10*B*). If the enveloping surface is itself folded, a second enveloping surface may be drawn to produce further simplification.

Harmonic and disharmonic systems. Where sets of folds in adjacent layers correspond with each other in wavelength, symmetry and general shape, the system is called *harmonic*. However, in many cases the wavelength and shape of folds in adjacent layers are quite different—often due to differences in physical properties or thicknesses (see section 10.3). Such fold systems are known as *disharmonic* (Figure 2.11; see also Figures 10.10, 10.11).

Conjugate and polyclinal systems. A pair of asymmetric folds with opposite sense of asymmetry, such that the axial surfaces dip towards each other, are termed *conjugate folds* (Figure 2.12*B*). A common type of conjugate fold is a *box fold* (Figure 2.12*A*) where the fold angles are approximately 90°, forming an almost rectangular structure. A *polyclinal* fold system is a more complex structure formed when the axial surfaces of adjacent folds have varying orientations (Figure 2.12*C*).

2.6 Folds in three dimensions

In the preceding sections, we have been discussing a geometrical classification of folds based essentially on the two-dimensional fold profile, ignoring the third dimension. Folds that maintain a constant profile are termed *cylindroidal folds*. Such folds may be generated by a line moving parallel to itself so that the fold surface produced contains a set of parallel lines. *Non-cylindroidal folds* are very common, and we shall discuss several special types.

Periclines, domes and basins. The term *pericline* is usually applied only to large-scale folds. A pericline is a fold whose amplitude decreases regularly to zero in both directions, so that the fold has precise limits in space (Figure 2.12*A*, *B*). Periclines may be anticlinal (antiformal) in which

A   box fold   B   conjugate kink bands

C   polyclinal folds

Figure 2.12 Conjugate and polyclinal folds. *A*. Box fold—a symmetric fold with four sets of hinges. *B*. Conjugate kink bands producing an asymmetric structure. *C*. Polyclinal folds with variable axial surfaces.

case they are termed *brachyanticlines*, or *brachysynclines* if synformal.

A *dome* is a special type of antiformal pericline where the dip is radial, that is, in plan view, the structure is close to circular. The synformal counterpart of a dome is called a *basin* (Figure 2.13*C, D*).

Culminations and depressions. In a surface showing non-cylindroidal folds, whether periclinal or more complex, the fold axes are generally curved and vary in height. Points of maximum elevation along antiformal fold axes are termed *culminations*, and points of minimum elevation along synformal fold axes are termed *depressions*.

Figure 2.13 Periclines, domes and basins. *A*. Oblique air photograph of part of a large anticlinal pericline in limestone, Asmari Mountain, Iran. Aerofilms Educational Series, sheet 22, courtesy Aerofilms Ltd.

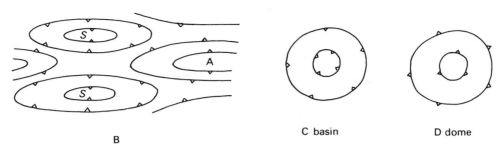

Figure 2.13 (contd.) Periclines, domes and basins. *B*. Plan view of synclinal (*S*) and anticlinal (*A*) periclines showing typical 'canoe'-shaped outcrops *C, D*. Plan views of a basin and a dome. Arrowheads indicate dip direction.

In many cases culminations will be domes and depressions basins.

Interference patterns and superimposed folds. Many complex fold systems are the result of 'interference' between two or more fold sets of simpler geometry. It is easy to imagine how the folding of a layer which already possesses a set of folds will produce a complicated three-dimensional shape (Figure 2.14*A*). Such structures are easily recognized by the outcrop patterns they produce. The type of outcrop pattern depends

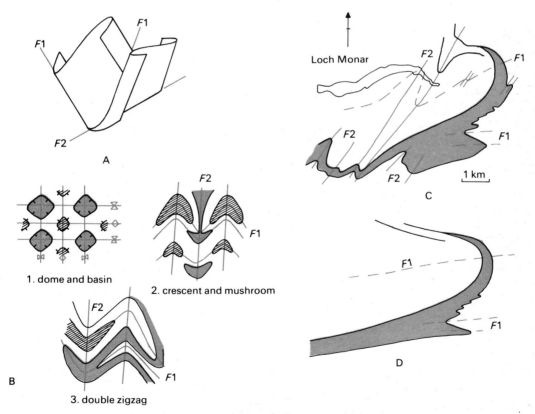

Figure 2.14 Superimposed folds. *A*. Refolding of an already folded layer, with folds *F*1, by a second fold *F*2 producing a complex three-dimensional shape. *B*. Three main types of outcrop pattern ('interference' pattern) produced by superimposition. Fold axial traces **shown in colour**. *C*. An example of an interference structure from Loch Monar, Ross-shire. *D*. The Loch Monar fold before *F*2 refolding. (*C* and *D* after Ramsay, J.G. (1967), *Folding and Fracturing of Rocks*, McGraw-Hill, New York. Fig. 10.20).

on the geometry of the two fold sets and on their relationship to each other. Three characteristic types of interference pattern are shown in Figure 2.14B—closed 'dome and basin' shapes, 'crescent and mushroom' or 'stirrup' shapes, and hooked or 'double zig-zag' shapes. In the latter two cases the order of superimposition is clear from the fact that the earlier folds and their axial traces are folded around the later fold axes. An example of a large-scale map interference pattern from the Loch Monar area of Scotland is shown in Figure 2.14C, D. The geometry of super-imposed 'flow' folds is discussed in more detail in section 10.4.

2.7 Relationship between faults, folds and ductile shears

At the beginning of the chapter it was pointed out that, whereas faults were the product of rather sudden brittle deformation, folds were formed by slow continuous changes under conditions of ductile deformation. However, the two types of process are not completely separate, as will be seen when we discuss the physical behaviour of rocks under deformation in section 7. For example, under certain conditions, folding may lead to fracture as deformation progresses. Moreover, layers of stronger material interbedded with weaker material may exhibit fracturing while the weaker material shows only ductile folding (Figure 2.15). Rocks also become more ductile at deeper levels in the crust, and a brittle fault at the surface may pass downwards

zone of ductile deformation

Figure 2.16 Simplified geometry of a shear zone. (Note that in most real shear zones, the marker layers would be curved through the zone rather than straight).

at depths of 10–20 km into a structure where the displacement between the two fault blocks is taken up by a type of ductile structure called a *shear zone*.

Shear zones. A shear zone is a zone of ductile deformation between two undeformed blocks that have moved relative to each other (Figure 2.16). There are no discrete fracture planes in an ideal shear zone, although in practice there is a complete gradation betweeen a fault zone and a shear zone, with intermediate stages being represented by faulted shear zones. The structure and mode of formation of shear zones are discussed in more detail in section 9.5.

Slides. The term *slide* was originally used for a fault that developed during folding. The type examples of such structures were low-angle thrusts or lags developed in association with large-scale recumbent folds and often themselves folded by subsequent deformation. In practice it is very difficult to distinguish between faults that precede the folding and those that develop contemporaneously. Some of the famous slides of the Scottish Highlands (the Sgurr Beag slide for example) are actually ductile shear zones developed under metamorphic conditions and associated with wide belts of intense deform-ation. Structural geologists generally now use the term 'slide' to describe structures of the latter type, where the slide either corresponds to a shear zone or to one or more faults within and genetically related to a shear zone.

Figure 2.15 Faults developed during folding. Ornamented beds are relatively competent and unornamented beds relatively incompetent.

Further reading

Fleuty, M.J. (1964) The description of folds. *Proc. geol. Ass.* **75**, 461–489.

Hobbs, B.E. *et al.* (1986) *An Outline of Structural Geology*, 2nd edn, John Wiley, New York.

Ramsay, J.G. and Huber, M.I. (1987) *The Techniques of Modern Structural Geology*, Vol. 2: *Folds and Fractures*, Academic Press, New York. See Chap. 2 for a comprehensive treatment of fold morphology.

See also section 10 references.

3 Foliation, lineation and fabric

Structures found in rocks from the deeper levels of orogenic belts are characteristically different from those found at higher levels in the crust. The difference is due mainly to the effect of the increased temperature and confining pressure in these regions, which increases the *ductility* of the rocks (see section 7), and to strong compression leading to intense and repeated folding.

Three important observations may be made concerning the structures at deeper crustal levels: *folding* is the typical mode of deformation rather than faulting; sets of *new planar surfaces* (cleavage, schistosity, etc.) are commonly developed; and *pervasive recrystallization* under compression results in the internal rearrangement of the rock texture producing a new *'fabric'*, or structural texture.

3.1 Foliation

Definition and meaning. A *foliation* is a set of new planar surfaces produced in a rock as a result of deformation. Foliation is a general term covering different kinds of structure produced in different ways.

Slaty cleavage, schistosity, gneissose banding and sets of *closely-spaced fractures* are all examples of foliations. Many rocks exhibit several generations of foliation which are distinguished chronologically (S1, S2, S3, etc.). Earlier foliations are deformed and cut by later foliations, enabling the structural history of the rocks to be established. The bedding is usually the first recognizable planar surface, and is sometimes designated SS or S0. Sometimes,

however, the bedding is obliterated by deformation and the first visible planar surfaces are tectonic; or the origin of the first set of surfaces may be indeterminate. In many areas there is a foliation parallel to bedding, often termed a *bedding foliation*. This may arise either as a result of the load pressure of overlying strata, or through deformation associated with folding, in which case the foliation will locally cut the bedding in fold hinges (Figure 3.4B).

Types of foliation. The nomenclature of the various types of foliation is rather confusing. This reflects the fact that the origin of such structures as slaty cleavage and gneissose banding, for example, was not fully understood until relatively recently. The term 'cleavage' itself embraces structures of various origin, the only common factor being the *fissility* which allows the rock to be split along the foliation planes. The main types of cleavage are *slaty cleavage*, the very pervasive cleavage found in typical roofing slates; *fracture cleavage*, a set of closely-spaced fractures; *crenulation cleavage*, a planar structure produced by microfolding; and *solution cleavage*, in which differential solution has produced a compositional banding.

Foliations may either be pervasive, or *penetrative*, affecting all parts of the rock, or they may occur at intervals separated by unaffected rock. *Non-penetrative cleavage* of this type is often termed *spaced cleavage*. Whether or not a foliation is regarded as penetrative depends on the scale of observation; penetrative foliation at outcrop or hand-specimen scale may appear as discrete planes in thin section. *Schistosity* is a foliation produced by parallel alignment of tabular or elongate minerals, such as micas or hornblende, in rocks that have undergone more intense metamorphic recrystallization. Another type of foliation is produced by compositional layering, similar to bedding but of metamorphic or deformational origin. *Gneissose banding* or *gneissosity* is an example of such a structure.

Slaty cleavage. This type of cleavage is best shown in fine-grained rocks such as mudstones that have been deformed under very low-grade

Figure 3.1 Examples of rock cleavage. *A.* Axial-plane cleavage in folded sandstones of the South Stack series, Rhoscolyn, Anglesey. *B.* Crenulation cleavage in finely laminated gneisses. The crenulations die out downwards into a larger-wavelength open fold. Härmanö, Orust, SW Sweden. *C.* Photomicrograph showing the development of crenulation cleavage in slate. The cleavage 'planes' correspond to zones of superimposed short limbs of asymmetric crenulations which are enriched in micas, etc., relative to quartz, thus forming darker layers. From photograph by W.D. Means (1977) in *Atlas of Rock Cleavage*, eds. B.M. Bayly, G.J. Borradaile and C. McA. Powell, University of Tasmania. *D.* Shape fabric foliation produced by the alignment of flattened and elongated pebbles in a deformed conglomerate. Stevenson Lake, Manitoba (photograph by I.F. Ermanovics).

metamorphism. Consequently the nature of the internal changes in the rock that have produced this penetrative fissility is not usually obvious at outcrop or in hand specimen. Under the microscope, however, the nature of the cleavage becomes much clearer. The cleavage planes are then seen to be due partly to the parallel orientation of flaky minerals such as muscovite and clay minerals, and partly to the parallel arrangement of tabular or lensoid aggregates of particles (cf. *shape fabric foliation*).

The origin of this structure becomes clear when we examine slates containing deformed objects of known initial shape, such as fossils, or reduction spots. The plane of the cleavage is seen to correspond to the plane of flattening of such deformed objects, which leads us to conclude that the parallel orientation of grain aggregates and tabular minerals is due to intense compression of the rock in a direction perpendicular to the plane of the cleavage. This compression has resulted in the rotation of previously-formed minerals and has also controlled the growth of new minerals, causing them to be aligned in the direction of the cleavage. Slaty cleavage appears to form only when a suitable rock has been compressed by about 30% of its initial length.

Fracture cleavage. As the name suggests, a fracture cleavage consists of parallel, closely-spaced fractures. Fracture cleavage is usually easy to distinguish from slaty cleavage because it consists of discrete planes separated by slabs of uncleaved rock, called *microlithons*. Displacement of the rock across these planes may often be visible in thin section, showing that the planes are microfaults (Figure 3.1*A*). This type of cleavage is formed under brittle conditions at low temperatures and is typical of deformed relatively strong rocks, e.g. sandstones and limestones.

Fracture cleavage may accompany slaty cleavage, either in the same rock or in adjoining layers, and in some cases displacements may take place on previously formed slaty cleavage planes. This has caused some confusion in the past over the origin of slaty cleavage.

Fractures may be compressional in origin (shear fractures) or extensional, in which case they will generally be filled by quartz or calcite (e.g. Figure 4.2). Certain examples of extensional fracture cleavage are thought to be due to *hydraulic fracturing* by water forced out of the sediments by load pressure.

Crenulation cleavage. This type of cleavage is caused, as the name suggests, by small-scale folding (crenulation) of very thin layers or laminations within a rock. If the axial surfaces of such crenulations are closely-spaced and parallel, they produce a marked foliation (Figure 3.2). This foliation is often enhanced by selective recrystallization leading to a concentration of certain constituents in layers. Thus micas, for example, may become concentrated in one set of limbs of asymmetric crenulations or in both sets of limbs of symmetric crenulations, as a result of the migration of quartz or calcite into limbs or hinges (Figure 3.1*C*). The combination of compositional banding and parallel orientation of platy minerals provides planar weaknesses that can impart a strong fissility to the rock.

Crenulation cleavage is very commonly

A asymmetric   axial surface traces   axial surface traces   B symmetric

Figure 3.2 Crenulation cleavage. *A*, asymmetric; *B*, symmetric. The cleavage is parallel to the axial surface traces (colour) of the crenulations.

associated with later deformation of rocks that possess a strong cleavage or schistosity as a result of an earlier deformation. The early foliation planes provide a well-laminated structure which assists the crenulation process.

Previously undeformed rocks may also possess a crenulation cleavage in suitable lithologies— generally very finely laminated shales. If the crenulations are on a sufficiently small scale, the resulting cleavage becomes indistinguishable from other types of slaty cleavage in hand specimen.

Schistosity. With increasing metamorphic grade, slates are transformed to schists by an increase in the size of the newly formed metamorphic minerals. In slates, the aligned flaky minerals that produce the slaty cleavage are invisible to the naked eye, whereas in schists, the individual tabular crystals of mica, hornblende, etc. are large enough to be visible in hand specimen. A foliation marked by the parallel orientation of such tabular minerals in a metamorphic rock with a sufficiently coarse grain size is called a *schistosity*.

A schistosity can be produced directly from a slaty cleavage merely by a coarsening of the grain-size, consequent on an increase in temperature. Crenulation cleavage may also pass into schistosity as a result of grain-size coarsening. Many schistose rocks show a combination of mineral alignment (true schistosity) and a tabular or lensoid arrangement, similar to that seen in many slates, produced by compression, but on a larger scale.

Shape fabric foliations. An important effect of deformation in many rocks is the parallel align-ment of deformed objects within the rock body. These objects may be grains, or grain aggregates (pebbles, ooliths, fossils, etc.). The structure produced by a set of parallel, dimensionally-oriented shapes is called a *shape fabric* (see section 3.4) and a planar shape fabric is a type of foliation (Figure 3.1D).

New compositional layering. A parallel banding produced by alternating layers of different composition is an important feature of many deformed metamorphic rocks, and can arise in several different ways. We have already discussed how the migration of certain constituents during the formation of a crenulation cleavage can produce such a banding. Preferential redistribution of minerals in a metamorphic rock is termed *metamorphic segregation* or *metamorphic differentiation* and is very important in producing compositional layering in deformed rocks. A common form of segregation in low-grade or even un-metamorphosed rocks is caused by *pressure solution* and the resulting cleavage is often termed 'solution cleavage' (Figure 3.3A). The solution and accompanying deposition appear to be part of a diffusion process which takes place by means of a grain-boundary fluid phase. Solution occurs on grain or layer boundaries that are perpendicular to the direction of greatest compression, and deposition takes place on surfaces that are perpendicular to the direction of extension or least compression. In its most common form, this structure consists of alternating lighter and darker bands, where the darker bands are produced by the removal by solution of calcite or quartz (Figure 3.3B).

Figure 3.3 Compositional layering produced by deformation. A_1. Solution cleavage in synform affecting Devonian slates, Tor Cross, Devon. The light-coloured silt layers are cut by a penetrative cleavage defined by seams enriched in dark material presumably due to the solution and redeposition of the light material elsewhere. A_2. Close-up of A_1. B. Photomicrograph of A showing dark layers where material is thought to have been dissolved. Bedding is clearly shown dipping steeply to the right (scale bar 2 mm). A, B from photograph by H.R. Burger (1977) in *Atlas of Rock Cleavage*, eds. B.M. Bayly, G.J. Borradaile and C. McA. Powell, University of Tasmania. C. Gneissose banding produced by deformation. The three photographs show three stages in the deformation of a coarse-gained leucogabbro from the Archaean of West Greenland. In C_1 the mafic areas are easily recognizable although generally elongate and recrystallized. In C_2 the rock has been transformed into a gneiss with alternating lensoid mafic and felsic components. In C_3 further deformation has resulted in a striped rock whose origins could not be determined from this outcrop alone. From Bridgwater, D., Keto, L., McGregor, V.R. and Myers, J.S. (1976) in *Geology of Greenland*, eds. A. Escher and W.S. Watt, Geological Survey of Greenland, Figs. 48–50.

Compositional layering is a characteristic feature of most gneisses and is often termed *gneissosity* or *gneissose banding*. Gneisses are coarse-grained metamorphic rocks, typically quartzo-feldspathic in composition. The distinction between schists and gneisses is not clear-cut, and individual geologists have their own preferences as to where the dividing line should be drawn.

The origin of the compositional layering in gneisses has occasioned much debate in the past and it is now clear that several different processes are involved. Many gneisses with a layered or lensoid structure (*augen gneisses*) have clearly originated as coarse-grained plutonic rocks, often rather variable in composition, with abundant cross-cutting veins. Under severe compression involving very high strains, such rocks become layered, due partly to the flattening and elongation of large crystals, and partly to the rotation of veins and other heterogeneities into the plane of flattening (Figure 3.3C). Metamorphic segregation is also important in the production of gneissose banding, either by enhancing a deformational layering or by producing a new layering by pressure solution.

Intensely deformed gneisses of sedimentary origin (*paragneisses*) when they are derived from sediments of mixed composition, e.g. greywackes or arkoses, are often very difficult to distinguish from those of igneous origin (*orthogneisses*). It cannot be assumed that the presence of a compositional banding necessarily indicates a sedimentary origin.

Relationship between folds and foliations. The deformation that is responsible for the formation of a foliation will normally also produce folds in a suitably layered rock. Since the two structures are produced by the same compression, they generally bear a simple geometrical relationship to each other. A foliation that corresponds to a plane of flattening is parallel or sub-parallel to the axial surface of a related fold (Figure 3.4A) or forms a fan-shaped structure arranged symmetrically about the axial surface (Figure 3.4C, D). Under intense deformation, tight to isoclinal folds are produced whose limbs

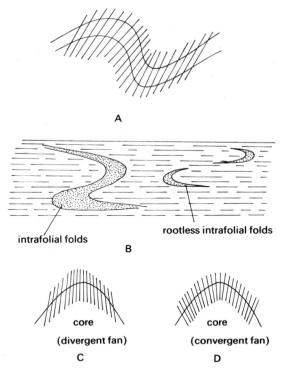

Figure 3.4 Relationship between folds and foliation. *A.* A foliation developed during folding is often sub-parallel to the fold axial surface. *B.* Intense deformation may cause thinning and transposition of fold limbs into the plane of the foliation ultimately causing the limb to disappear (see text). Types of folds thus formed are intrafolial and rootless intrafolial folds. *C, D.* Foliation developed during folding often has a simple fan-shaped arrangement about the axial surface, *C*, divergent fan; *D*, convergent fan.

may become thinned and modified until they are indistinguishable from the foliation, and only the fold hinges are clearly definable. Such folds are termed *intrafolial folds* (Figure 3.4B). Where the hinges are completely detached from the limbs, the folds are called *rootless intrafolial folds*.

The association between folds and foliations conveys valuable information about the conditions of deformation of a rock and is discussed more fully in section 10.

3.2 Lineation

A *lineation* is the linear counterpart of a foliation—that is, a set of linear structures produced in a rock as a result of deformation.

There are many types of lineation, and one of the problems facing an inexperienced geologist in the field is to distinguish significant from insignificant lineations. Since any two planar surfaces intersect in a line, the more planar surfaces there are in an exposure, the more potential lineations there will be. Common examples of lineation are *slickenside striations, crenulation fold axes, elongated pebbles* in a deformed conglomerate, *lines of intersection* of bedding and cleavage in slate, and *alignment of elongate minerals*.

A convenient subdivision of lineations can be made into the following groups:

(i) lineations indicating the direction of movement along a surface (e.g. slickenside striations)
(ii) axes of parallel crenulations or small-scale folds
(iii) dimensional elongation of a set of deformed objects such as pebbles, ooids, megablasts, etc.
(iv) parallel orientation of elongate minerals (mineral lineations)
(v) intersections of sets of planes (intersection lineations).

Lineations indicating movement direction. Fault surfaces showing slickensides (section 1.3) commonly exhibit grooves or striations indicating the direction of relative movement (Figure 3.5*A*). A similar type of lineation is often found on bedding surfaces involved in *'flexural slip'* folding (see section 10.1) where successive layers have moved over one another as the folds tighten (Figure 3.5*B*). Such lineations usually make a large angle with the fold axis. Both these types of lineation lie in a particular surface and do not permeate the body of rock in which they are found. They may thus be termed non-penetrative.

Axes of parallel crenulations. Rocks that are finely laminated and affected by intense small-scale folding or crenulation exhibit a strongly developed linear structure due to the parallel fold hinges that permeate the rock (Figure 3.5*C*). Slaty or schistose rocks frequently exhibit two or more sets of such crenulations and crenulation

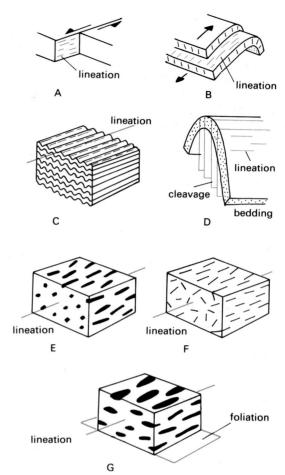

Figure 3.5 Types of lineation. *A.* Slickenside striation on a fault *B.* Striations on a bedding plane caused by flexural slip. *C.* Crenulation lineations. *D.* Intersection lineation between bedding and cleavage. *E.* Mineral elongation lineation. *F.* Mineral intersection lineation. *G.* Fabric that is partly planar and partly linear.

cleavages, and individual foliation surfaces may then contain two intersecting sets of crenulation lineations forming a small-scale interference pattern.

Intersections of planes. One of the commonest types of lineation is formed when two sets of planar structures intersect. Strongly-developed lineations are often found at the intersection of bedding and cleavage (Figure 3.5*D*) or of two foliations—a schistosity and a crenulation cleavage for example. Such *intersection lineations* are very often parallel to local fold axes and crenulation lineations that belong to the same episode of deformation.

Great care must be taken in the field to determine the origin of a lineation before it is recorded and measured, since the structural significance of different lineations may be quite different. Many insignificant lineations may be observed at an exposure, caused for example by the intersection of a foliation on a random joint surface, or on the surface of the exposure itself, and these should be ignored. Crenulation lineations and intersection lineations parallel to the local fold axes were often called 'b'-lineations, and lineations parallel to the movement direction were called 'a'-lineations in the older literature, but these terms are no longer widely used and are not recommended.

Dimensional elongation. An important type of lineation is formed by the parallel alignment of a set of objects within a rock body that possess a linear elongation as a result of deformation. There is a wide variety of objects that may define such an elongation lineation. Obvious examples are pebbles, ooids and spherulites, but individual grains or grain aggregates, or indeed any kind of heterogeneity within the rock, may acquire an elongate shape as a result of deformation and contribute to an elongation lineation. In strongly deformed rocks, such a lineation may pervade the whole rock and consist of many different types of oriented feature. Figure 3.6A shows an example of dimensional elongation in an intensely deformed pillow lava.

Rodding and mullion structure. Many highly deformed rocks possess a rodding structure whose origin is not immediately apparent. Rods may be formed by the hinges of crenulated quartz veins, or may represent extremely elongate pebbles. In rocks that have been subjected to very high strains, pebbles that are only a few centimetres across may measure one or more metres long, in which case the true nature of the lineation may not be apparent at first sight. A rodded structure with dimensions of tens of centimetres

across is often termed *mullion structure*. A common type of mullion is formed at the interface between competent and incompetent layers (sandstone and shale for instance) and results partly from the folding of the interface and partly from the effect of bedding–foliation intersection (Figure 3.6B). Rodding or mullion structure may represent an elongation lineation in some cases (cf. boudinage—see section 3.3), but each example should be carefully examined to determine its true nature.

Mineral orientation. Parallel linear orientation of crystals is very common in deformed metamorphic rocks. Lineations of this type are usually called *mineral lineations* (Figure (3.5E). The parallel alignment of the crystals forming the lineation may be due to rotation into a favoured attitude as a result of the deformation, or it may be due to the effect of recrystallization under stress, when certain crystallographic orientations are encouraged and others suppressed. Minerals that grow preferentially in a particular direction under stress are said to show *growth anisotropy*.

Mineral lineations may be caused by the alignment of crystals with an elongate habit, such as hornblende for example (Figure 3.5E), but they may also be caused by the alignment of minerals with a platy habit, e.g. micas, if they are arranged in such a way as to form an intersection lineation (Figure 3.5F). In the latter case, they may be arranged randomly in the plane perpendicular to the lineation.

Many mineral lineations are associated with foliations, particularly where the latter are also formed by mineral orientation, or, as is commonly the case, by a combination of dimensional and crystallographic planar orientation (Figure 3.5G). In such cases, the planar and linear elements are both aspects of the same three-dimensional geometry which is related to the shape of the *'strain ellipsoid'* (section 6.3). We may envisage a continuous progression from a purely linear structure with no planar element through a variety of structures produced by combinations of elongation and flattening to a purely planar structure with no linear element.

Figure 3.6 Examples of linear structure *A*. Strongly developed rodding lineation in Lewisian siliceous schists from the Loch Maree Group, Gairloch, Ross-shire. *B*. Mullion structure in vertical bedding surface between sandstone and slate—the slate has been mostly removed by erosion. North Eifel, Germany. From Hobbs *et al.* (1986), Fig. 6.9.

3.3 Boudinage

When relatively strong layers of rock become stretched and elongated during deformation, they may separate into blocks (Figure 3.8) or form lensoid or pillow-shaped structures separated by narrow 'necks'. Such structures are called *boudins* (or *'pull-apart'* structure) and the process of elongation which produced them is called *boudinage*. Where the separation is incomplete and the layers show a narrowing or '*necking*', the structure is often termed *pinch-and-swell*. Boudins may either be linear and form an elongation lineation (Figure 3.7*A*) or they may form a *chocolate-tablet structure* if extension has occurred in both directions within the plane of the layer (Figure 3.7*B*).

Boudins characteristically form in relatively thin layers (usually up to about 1 m in thickness) that are stronger and more competent than the enclosing rock, which therefore tends to stretch in a ductile fashion and flow into the spaces between the boudins. Many boudins, particularly of the 'pull-apart' type, are separated by veins of quartz or calcite or, in high-grade metamorphic rock, pegmatitic material.

Boudins are particularly useful as indicators

Figure 3.8 Folded boudinage shown by thin dykes in Precambrian gneisses at Tromøy, S. Norway.

of directions of extension in deformed rocks. The relationship between boudinage and folding during progressive deformation is discussed in section 6.8.

3.4 Fabric

The *fabric* of a rock body is the geometrical arrangement of all the structural elements within the body. It can be regarded as the structural 'texture' of the rock. Only structural elements observed on a relatively small scale (hand-specimen or microscopic size) are normally considered as fabric, not large-scale structures. Thus fabric study effectively involves structural analysis at grain-size level.

A fabric is made up of a number of *fabric elements* each consisting of a group of geometric features of the same kind. These elements are usually either *planar* or *linear*. Planar fabric elements include the small-scale expressions of a foliation or lineation, e.g. dimensional or crystallographic orientation of grains and grain-aggregates. They also include various kinds of planar discontinuities such as grain-boundaries, twin planes and dislocations within crystals. Fabric elements usually produce a planar or linear orientation, but it is possible to have '*random fabrics*' consisting of randomly-oriented fabric elements. A fabric composed of dimensionally oriented objects (grains or grain-aggregates) is known as a *shape fabric*.

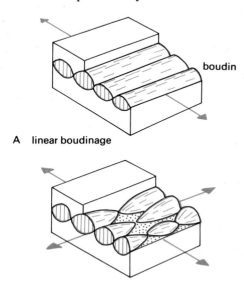

A   **linear boudinage**

B   **chocolate-tablet boudinage**

boudin

Figure 3.7 Boudinage. *A*. Linear boudinage. *B*. 'Chocolate-tablet' boudinage produced by two directions of extension.

In exactly the same way as foliations and lineations, the fabric of a rock expresses the strain geometry and reflects the way in which the rock has accommodated itself to deformation by changes in the shape, pattern and orientation of the grain network.

Homogeneity of fabrics. The fabric of a rock is said to be *homogeneous* if any two parts, similarly oriented, show identical structure. The term is usually applied to a particular fabric element (a foliation, for example). The scale of observation is important, since the fabric element in question may be homogeneous at hand-specimen scale but heterogeneous at thin-section scale. *Heterogeneous fabrics* can be divided into homogeneous *domains* in order to simplify the analysis (Figure 3.9).

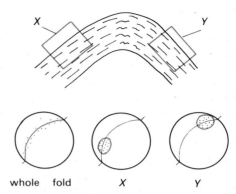

Figure 3.9 Homogeneous domains (X, Y) in a heterogeneous fabric. Stereograms show poles to planar fabric over the whole fold and in domains X and Y.

Figure 3.10 Microfabric at crystal scale. *A.* Deformation bands in naturally deformed quartz. *B.* Kink boundaries in naturally deformed biotite. Both *A* and *B* from Arunta complex, central Australia; scale mark, 0.3 mm; crossed polars. From Hobbs *et al.* (1986), Fig. 2.11. *C.* Flattened grain texture resulting from intragranular gliding and twinning in experimentally deformed calcite. C_1, undeformed Carrera marble; C_2, shortened by 50% at $400\,°C$ and 1.5 kbar confining pressure at 10^{-5}/s strain rate. Plane polars. Photographs by E.H. Rutter and N.D. Shaw (1977) in *Atlas of Rock Cleavage*, eds. B.M. Bayly, G.J. Borradaile and C. McA. Powell, University of Tasmania.

Microfabric elements. Detailed study of how deformation affects individual grains, carried out particularly with the use of the electron microscope, has shown that deformation on the submicroscopic scale is very heterogeneous and takes place by displacements along a series of dislocations or discontinuities within grains and on grain boundaries. The processes involved in the production of the crystal microfabric are discussed in section 7.11, and readers are also referred to the textbook by Hobbs *et al.* (1986) for a comprehensive treatment of this subject.

There are a number of different types of discontinuity at individual crystal scale. In addition to the grain boundaries, these discontinuities consist principally of *crystal defects*, the nature of which is controlled by the molecular structure of the crystal. Planar defects are normally parallel to one of the crystallographic planes in a crystal and allow displacements to take place within the crystal. These displacements enable the crystal to change its shape to accommodate to the deformation. Many crystals show planar defects, usually of limited extent, across which there is a very small displacement of the crystal structure. These defects are called *stacking faults* and are important in the propagation of dislocations that may subsequently form more pervasive deformation planes.

Sub-grain boundaries are planar defects within grains separating regions of slightly different lattice orientation and are visible as a small change in extinction angle (Figure 3.10*A*, *B*). Such boundaries cause the well-known phenomenon of *undulose extinction* in quartz.

Deformation bands are narrow planar zones containing material that has deformed differently from the adjoining parts of the crystal, either by a small change in lattice orientation, or by a more complicated series of changes. *Deformation lamellae* are a special type of deformation band that possess a uniform structure but exhibit a different refractive index from the host crystal. Another type of deformation band is caused by *deformation twinning* (Figure 3.10*B*) which is common in many crystals, especially calcite. This structure is produced by a slight rotation of the crystal lattice, in a homogeneous manner, between the twin planes.

Further reading

Hobbs, B.E. *et al.* (1986) *An Outline of Structural Geology*, 2nd edn, John Wiley, New York. Contains an excellent section on microfabrics.

Turner, F.J. and Weiss, L.E. (1963) *Structural Analysis of Metamorphic Tectonites*, McGraw-Hill, New York. This is still the most comprehensive descriptive work on tectonic fabrics, the geometrical analysis of structures and the use of stereographic projection in structural geology.

The following references should only be read after Part 2 of this book has been tackled.

Cosgrove, J.W. (1976) The formation of crenulation cleavage. *J. geol. Soc. Lond.* **132**, 155–178.

Gray, D.R. and Durney, D.W. (1979) Crenulation cleavage differentiation: implications of solution-deposition processes. *J. struct. Geol.* **1**, 73–80.

Platt, J.P. and Vissers, R.L.M. (1980) Extensional structures in anisotropic rocks. *J. struct. Geol.* **2**, 397–410. Discusses boudinage in the context of a general treatment of extensional structures.

Price, J.N. and Hancock, P.L. (1972) Development of fracture cleavage and kindred structures. *Proc. 24th int. geol. Congr.* Section 3, 584–592. Discusses the hydraulic fracturing mechanism.

Wood, D.S. (1974) Current views of the development of slaty cleavage. *Annual Review of Earth and Planetary Sciences* **2**, 369–401.

4 Igneous bodies

4.1 The significance of igneous bodies in structural geology

A detailed discussion and classification of igneous bodies is outside the scope of this book, but we must concern ourselves with certain aspects of igneous bodies that are directly relevant to structural geology. This relevance may arise in three ways. Firstly, an igneous body may contain structures (particularly foliation and lineation) that are caused by deformation, either during or after intrusion. Secondly, the shape and orientation of many igneous bodies is a direct result of the pre-existing structure of the country rock into which they have been emplaced. Lastly, and most importantly, the forces within the Earth's crust that give rise to deformation in solid rocks also control the way that magma is emplaced and, to some extent, determine the geometry of the resulting igneous bodies.

4.2 Structures found within igneous bodies

It is often difficult in practice to distinguish between structures formed as a result of the movement of magma during emplacement (*igneous structures*) and those structures formed after emplacement as a result of deformation of the solid rock. A similar difficulty in distinguishing some sedimentary structures was discussed in the introduction. The commonest type of igneous structure is a *flow foliation* caused by the alignment, during the flow of the magma, of tabular minerals that had formed before the final consolidation of the magma. A *flow lineation* may be formed in the same way by the alignment of elongate crystals, but is less common. Sometimes a *flow banding* is formed in very viscous magmas such as rhyolite. This banding may locally be involved in intricate folding rather similar to the folds formed in solid rock under highly ductile conditions. Another type of banding or lamination is produced by *gravitational settling* within a magma chamber.

Many large granite bodies exhibit a foliation that is parallel to their margins and decreases in intensity towards the interior of the body. Studies of the shapes of xenoliths, for example, show that this decrease in intensity is related to a regular pattern of deformation attributed to the 'ballooning' effect of later pulses of magma on earlier, partly or wholly consolidated, material. However, other examples of this phenomenon have been attributed to deformation in the solid state brought about by upward diapiric flow under gravitational pressure (see section 12.4).

The most reliable method of distinguishing true igneous structures is to look for evidence of undeformed igneous textures. An igneous rock deformed in the solid state, on the other hand, will tend to show a penetrative fabric.

4.3 Structural classification of igneous bodies

We shall consider only *intrusive* igneous bodies, since these are of most interest to the structural geologist. Intrusions are usually divided somewhat arbitrarily into major and minor. The major intrusions or *plutons* are large bodies of various shapes and sizes comprising many cubic kilometres of igneous rock. Minor intrusions are normally sheet-like or pipe-like in form where at least one dimension is of the order of metres rather than kilometres.

Sheet-like bodies. The commonest types of minor intrusion are *dykes* and *sills*. These are essentially parallel-sided, sheet-like bodies, which differ only in that dykes are largely discordant with the

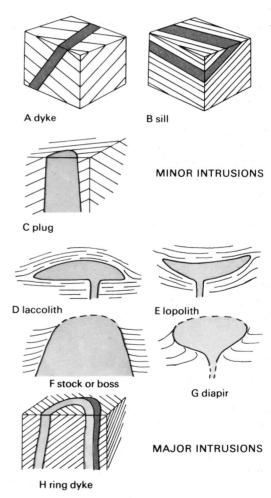

A dyke   B sill

MINOR INTRUSIONS

C plug

D laccolith   E lopolith

F stock or boss   G diapir

MAJOR INTRUSIONS

H ring dyke

Figure 4.1 Classification of intrusive igneous bodies based on form and structural relationships.

country rock structure and normally steeply-dipping (Figure 4.1*A*), whereas sills are largely concordant, and typically sub-horizontal or gently-dipping (Figure 4.1*B*). Dykes often occur as a set of related bodies termed a *dyke swarm* which may have a sub-parallel, radial or some other arrangement. *Cone-sheets* are a special type of dyke swarm where the dykes are arranged in a set of conical surfaces inclined towards a central point (Figure 11.5*B, C*).

Sheet-like minor intrusions are normally emplaced along fractures. The pattern of these fractures, and how they relate to local and regional stress fields, are of great interest to the structural geologist and are discussed more fully in section 11.

Other small-scale bodies. *Plugs* (Figure 4.1*C*) are pipe-like bodies of about 100–1000 m across that commonly fill the vent or neck of a volcano. Many eroded volcanoes are represented at depth by such bodies. The name *vein* is given to a small body (centimetres to metres in width) that is either sheet-like or quite irregular in form. Veins of igneous rock are usually *apophyses* or branches leading from a larger body although, in certain metamorphic regions, granitic vein complexes are known that have no obvious source and therefore have been attributed to mobilization or fractional melting of country rock. An important class of vein is filled by minerals deposited from aqueous solution. Such veins are extremely common, particularly those composed of quartz and calcite, and are a valuable source of ore minerals. From a structural point of view, veins are useful in indicating that fractures are dilational, that is the wall rocks have been moved aside to allow the vein material to form (Figure 4.2, and see also Figure 11.1). Dilational veins can therefore be used to indicate directions of relative extension in the rock.

Major intrusions. Large intrusive bodies or *plutons* may be either broadly concordant or broadly discordant. The concordant bodies, which may exhibit considerable discordance locally, include *laccoliths* and *lopoliths*. Laccoliths make room for themselves by arching up the strata above the intrusion, to form a lensoid shape (Figure 4.1*D*). Lopoliths are less common. They are accommodated by a sagging or down-bending of the strata beneath to form a saucer-shaped intrusion (Figure 4.1*E*). The type example of such a body is the Bushveld complex of South Africa, one of the largest igneous bodies in the world, with an outcrop area of more than $67\,000\,km^2$.

Discordant plutons may be quite irregular in form but very often they are roughly circular or elliptical in plan with steeply-dipping walls. The

Figure 4.2 Dilational quartz-filled vein fissures in sandstone, North Cornwall. The en-echelon veins mark a dextral shear zone which has rotated the initially planar vein into a sigmoidal shape. From Ramsay, J.G. (1980) *J. struct. Geol.* **2**, 83–99, Fig. 2*b*.

latter type are known as *stocks* or *bosses* (Figure 4.1*F*). The arched top of such bodies forms a *dome*. It is frequently difficult to establish the geometry of such large bodies at depth, and in the past many bodies were assumed to continue downwards indefinitely, until gravity surveys or more detailed mapping established their sheet-like or laccolithic form. Many igneous bodies are now regarded as *diapirs* whose shape swells upwards and outwards from a relatively narrow 'neck' (Figure 4.1*G*). Such bodies are thought to rise through the crust like very large blobs of liquid with an inverted tear-drop or pear shape.

A special class of pluton consists of one or more bodies with a ring-shaped cross-section and steeply-dipping walls. These bodies are called *ring dykes* (Figure 4.1*H*) and are often associated with stocks, radial dyke-swarms and cone-sheets in a 'central' igneous complex. There are several well-known examples of such complexes in the Tertiary igneous province of north-western Scotland, including Ardnamurchan, Mull and Skye.

The largest major intrusions are known as *batholiths*. Such bodies are usually elongate and may be many hundreds of kilometres long. Where they are well known, they have been shown to be composed of a number of individual plutons of various forms. Batholiths are typically granitic in composition although they may include a wide range of other rock types as well.

Further reading

Holmes, A. (1978) *Principles of Physical Geology*, 3rd edn, Nelson, London, 162–187.
See also section 11 references.

Part 2: Deformation—how structures are formed

In part 2, structures are discussed and explained in terms of processes of formation (i.e. *deformation*). We start with the theoretical treatment of deformation in terms of a system of causal forces that act on a rock body to produce stresses, and of the geometrical changes resulting from these stresses which constitute the *strain*, or deformation. The *behaviour of materials* in response to stress is then discussed, together with the various physical controls, such as temperature and confining pressure, that influence this response. We turn next to various methods of quantitative *determination of strain* in rocks, and finally, in sections 9–12, we deal with the deformation mechanisms involved in *faulting, folding, emplacement of igneous intrusions* and *gravity-controlled structures*.

5 Stress

The deformation of a material is the process whereby physical changes are produced in the material due to the action of applied forces. The forces that act on the rocks of the Earth's crust arise in various ways. The most important of these forces are due to *gravity* and to the *relative movements* of large rock masses in the crust and upper mantle. Since gravitational force is proportional to mass, the weight of an overlying column of rock constitutes a very significant force on rocks at depth in the crust.

The forces acting on a portion of rock produce a set of *stresses*, and the amount of deformation caused by these stresses is measured by the change in dimensions of the body. This change may consist of a change in shape, or volume, or both shape and volume (Figure 5.1) and constitutes the *strain* (see section 6).

5.1 Force and stress

In order to understand the concept of stress, we must first define 'force': a *force* is the product of a mass and its acceleration. Force is a vector quantity possessing both *amount* and *direction*, and can be represented by a line whose length specifies the amount and whose orientation specifies the orientation of the force. The sense of direction may be indicated by an arrow.

Resolution of forces. Figure 5.2A shows how a force F may be resolved into two components $F1$ and $F2$ at right angles and, conversely (Figure 5.2B), how any two forces $F1$ and $F2$ may be represented by their resultant F. By extending

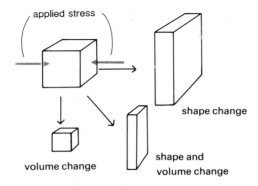

Figure 5.1 Effects of stress on a cube: a change in shape and/or volume.

Figure 5.2 Resolution of forces. *A*. Force F resolved into two components $F1$ and $F2$. *B*. two forces $F1$ and $F2$ represented by resultant F (see text).

32

this principle, it is clear that any system of forces acting at a point can be represented by a single resultant force.

Definition of stress. In rock deformation, we usually neglect any overall acceleration of a body and treat the system of forces as closed—i.e. opposing forces cancel out. This situation is governed by *Newton's Third Law of Motion* which states that 'for a body at rest or in uniform motion, to every action there is an equal and opposite reaction.'

We can now define stress: a *stress* is a pair of equal and opposite forces acting on unit area of a body. Thus a stress results from a force acting on a surface (either real or imaginary) surrounding or within a body, and comprises both the force and the reaction of the material on the other side of the surface. The magnitude of the stress depends on the magnitude of the force and on the size of the surface area on which it acts. Thus stress = force/area.

A body force such as *gravity* can give rise to a stress measured by calculating its effect across a surface. Gravity makes an important contribution to the stress field governing the formation of folds and faults.

Units of measurement. The standard SI unit for *force* is the *newton* (N) = 1 kilogram metre per square second (1 kg m s^{-2}). The SI unit for both pressure and *stress* is the *pascal* (Pa) = 1 newton per square metre (1 N m^{-2}). A more commonly used unit is the *bar* or *kilobar*. 1 bar = 10^5 pascals = 0.1 MPa. The dimensions of stress are thus (mass) (length^{-1}) (time^{-2}).

5.2 Normal stress and shear stress

A force F acting on a body can be resolved into a *normal stress* acting perpendicular to a surface within the body and a *shear stress* acting parallel to the surface (Figure 5.3A). Normal stresses are conventionally given the symbol σ (sigma) and shear stresses τ (tau). In three dimensions it can be seen that τ can be resolved into two further components τ_1 and τ_2 at right angles to each other. Thus we have converted the force F into

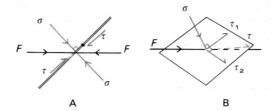

Figure 5.3 *A*. Normal stress σ perpendicular to the plane and shear stress τ parallel to the plane produced by opposed forces F acting on a plane (in two dimensions). *B*. In three dimensions, shear stress τ can be further resolved into τ_1 and τ_2 at right angles giving three stresses, all mutually at right angles, resulting from the forces F.

three mutually perpendicular stresses (Figure 5.3B). Note that stresses cannot be resolved in the same way as forces—they have to be converted into forces by multiplying them by the area on which they act.

Applications to geological structures. The application of normal and shear stresses can be illustrated with reference to two simple geological examples: the stress at a fault plane (Figure 5.4A) and the stress at a bedding plane undergoing flexural slip folding (Figure 5.4B), resulting from opposed compressive forces F. Clearly the sense of fault displacement and bedding-plane slip can be predicted if the direction of the force is known, and vice versa.

Figure 5.4 Normal and shear stresses at a fault plane (*A*) and a bedding plane during flexural slip folding (*B*) produced by resolving opposed compressive forces F.

Figure 5.5 Stress components for an infinitesimal cube acted on by opposed compressive forces F (see text).

5.3 Stress at a 'point'—the stress components

In order to consider the state of stress at a point in three-dimensional space, we must imagine the effect of a system of forces on an infinitesimal (vanishingly small) cube (see Figure 5.5). The system of forces can be resolved into a single force F which acts at the centre of the cube. Since the cube is very small, we can consider the forces acting on each face of the cube to be equal to F. If we make the edges of the cube parallel to orthogonal axes x, y and z, the components of stress acting on the cube are shown in Figure 5.5. There are nine stress components, three on each face (see Figure 5.3B). Because the forces are equal and opposite, the stresses on opposite faces are identical. The nine components are

$$\begin{matrix} \sigma_x & \tau_{xy} & \tau_{xz} \\ \sigma_y & \tau_{yx} & \tau_{yz} \\ \sigma_z & \tau_{zx} & \tau_{zy} \end{matrix}$$

Since the definition of stress precludes any contribution from an overall rotation of the cube, opposing shear stresses about the axes x, y and z must balance—otherwise the cube would rotate about these axes. Thus $\tau_{xy} = \tau_{yx}, \tau_{xz} = \tau_{zy}$ and $\tau_{yz} = \tau_{zy}$, leaving six independent stress components, three normal stresses and three shear stresses, $\sigma_x \sigma_y \sigma_z$, and $\tau_{xy} \tau_{yz} \tau_{zx}$. Therefore, for an arbitrarily chosen set of orthogonal axes x, y and z, six independent quantities are necessary to specify completely the state of stress at a point.

5.4 Principal stresses and the stress axial cross

Rather than use arbitrary axes x, y and z, it is convenient to choose new axes a, b and c such

Figure 5.6 The stress axial cross (principal stress axes $\sigma_1 \geqslant \sigma_2 \geqslant \sigma_3$) see text.

that the shear stresses are zero. That is:

$$\tau_{ab} = \tau_{bc} = \tau_{ca} = 0$$

The three mutually perpendicular planes on which the shear stress is zero are called *principal stress planes*, and the normal stresses across them are called the *principal stress axes*. These are given the conventional notation σ_1, σ_2 and σ_3 (where $\sigma_1 \geqslant \sigma_2 \geqslant \sigma_3$) or *greatest*, *intermediate* and *least principal stresses*, respectively. A state of stress is specified completely by giving both the directions and the sizes of each of the three principal stresses.

The stress axial cross. The three mutually perpendicular stress axes are often termed the *stress axial cross* in which the lengths of the axes may be drawn proportional to the magnitudes of the principal stresses (Figure 5.6).

5.5 Stresses acting on a given plane

If the principal stresses are known, the stresses acting on any plane with known orientation can be calculated. The problem is easier to visualize in two dimensions.

Consider the stresses acting on a plane AB whose normal makes an angle θ (theta) with σ_1 in a two-dimensional stress field with principal stresses σ_1 and σ_2 (Figure 5.7). Since we cannot resolve stresses σ_1 and σ_2 we must convert these stresses to forces. Let the line AB represent unit length (one side of a square of unit area in three dimensions). Then $OA = \sin \theta$ and $OB = \cos \theta$. The forces acting along OA and OB are thus $\sigma_1 \cos \theta$ and $\sigma_2 \sin \theta$ respectively (from force $=$ stress $\times$ area). Resolving these forces per-

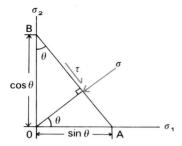

Figure 5.7 Normal and shear stresses on a plane inclined to the principal stress axes (in two dimensions). See text for explanation.

Figure 5.8 Planes of maximum shear stress (colour) make angles of 45° with the principal stress axes. There are three sets of planes intersecting in σ_3, σ_1 and σ_2.

pendicular and parallel to the plane AB, the normal stress σ and shear stress τ are as follows:

$$\sigma = \sigma_1 \cos^2 \theta + \sigma_2 \sin^2 \theta \qquad (5.1)$$

and

$$\tau = \sigma_1 \sin^2 \theta - \sigma_2 \cos^2 \theta \qquad (5.2)$$

Since

$$\cos^2 \theta = \tfrac{1}{2}(1 + \cos 2\theta)$$

and

$$\sin^2 \theta = \tfrac{1}{2}(1 - \cos 2\theta)$$

we can rewrite these equations as follows:

$$\sigma = \tfrac{1}{2}(\sigma_1 + \sigma_2) + \tfrac{1}{2}(\sigma_1 - \sigma_2)\cos 2\theta \qquad (5.3)$$

and

$$\tau = \tfrac{1}{2}(\sigma_1 - \sigma_2)\sin 2\theta \qquad (5.4)$$

Maximum shear stress. The value of τ in the last equation is a maximum when $2\theta = 90°$ and $\sin 2\theta = 1$. Thus the planes of maximum shear stress make an angle of 45° with σ_1 and σ_2 regardless of the values of σ_1 and σ_2.

In these positions,

$$\tau = \tfrac{1}{2}(\sigma_1 - \sigma_2) \qquad (5.5)$$

Stress in three dimensions. The geometry in three dimensions can be derived from the above by considering a plane of unit area making angles of θ_1, θ_2 and θ_3 with the three principal stress axes σ_1, σ_2 and σ_3. The normal stress on this plane is

$$\sigma = \sigma_1 \cos^2 \theta_1 + \sigma_2 \cos^2 \theta_2 + \sigma_3 \cos^2 \theta_3 \qquad (5.6)$$

and the shear stress is given by

$$\begin{aligned} \tau^2 &= (\sigma_1 - \sigma_2)^2 \cos^2 \theta_1 \cos^2 \theta_2 \\ &+ (\sigma_2 - \sigma_3)^2 \cos^2 \theta_2 \cos^2 \theta_3 \\ &+ (\sigma_3 - \sigma_1)^2 \cos^2 \theta_3 \cos^2 \theta_1 \end{aligned} \qquad (5.7)$$

The reader should refer to a more advanced textbook (e.g. Ramsay, 1967) for the derivation of the last equations.

Planes of maximum shear stress. From this equation, three sets of planes of maximum shear stress may be found making angles of 45° with each pair of the three principal stresses σ_1, σ_2 and σ_3 and intersecting the third (Figure 5.8).

Directions of maximum shear stress in a plane. If a plane makes an angle with all three principal stress axes (Figure 5.9), the direction of maximum shear stress in the plane is given by

$$\tan \alpha = \frac{\cos \theta_2}{\cos \theta_1 \cos \theta_3} \times$$
$$\left[\cos^2 \theta_2 - (1 - \cos^2 \theta_3)\frac{\sigma_3 - \sigma_1}{\sigma_2 - \sigma_3} \right] \qquad (5.8)$$

where, assuming for convenience that two of the principal stress axes are horizontal, α is the angle between the required direction and the horizontal in the plane and θ_1, θ_2 and θ_3 are the angles made by the plane with σ_1, σ_2 and σ_3 respectively.

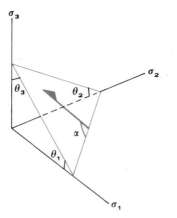

Figure 5.9 Shear stress on a plane inclined at angles θ_1, θ_2 and θ_3 to the principal stress axes. The shear direction is shown by the arrow and makes an angle α with the strike of the plane.

5.6 Hydrostatic and deviatoric stresses

Where the principal stresses are equal, the state of stress is said to be *hydrostatic*, i.e. it corresponds to the stress state of a fluid. It may be seen from equation (5.5) that the shear stress τ is zero in this situation. Hydrostatic stress will cause volume changes but not shape changes in a material.

In a system with unequal principal stresses σ_1, σ_2 and σ_3 it is convenient to recognize a mean stress P, which represents the *hydrostatic stress component*. Thus

$$P = (\sigma_1 + \sigma_2 + \sigma_3)/3 \qquad (5.9)$$

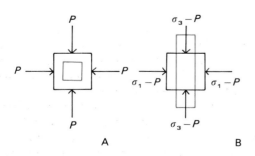

Figure 5.10 Effects of hydrostatic and deviatoric stresses (shown in two dimensions). *A*. Hydrostatic stress P causes a volume change. *B*. Deviatoric stresses $\sigma_1 - P$ and $\sigma_3 - P$ cause a shape change.

The remaining part of the stress system is referred to as the *deviatoric stress component*, which consists of three deviatoric stresses $\sigma_1 - P$, $\sigma_2 - P, \sigma_3 - P$. These deviatoric stresses measure the departure of the stress system from symmetry and control the extent of shape change or distortion in a body, whereas the hydrostatic stress component controls the change in volume (Figure 5.10). Stresses in rocks at depth that are hydrostatic and due solely to the weight of overlying rock are termed *lithostatic*. Note that lithostatic stress (or pressure) will not necessarily correspond to the mean stress, P.

5.7 Stress fields and stress trajectories

Until now we have been considering stress at a 'point', but normally stresses will vary throughout a rock body forming what is known as a *stress field*. Stress variation can be portrayed and analysed using *stress trajectories*, which are lines showing continuous variation in principal stress orientation from one point to another through a body. Two-dimensional stress trajectories of σ_1 and σ_2 are shown in Figure 5.11. Individual trajectories may be curved or bent but obviously the principal stresses must remain at right angles to each other at each point in the curve. Examples of the use of stress trajectories in analysing fault and dyke patterns are given in subsequent sections (see Figures 9.6, 11.5).

Combination of stress fields. Frequently, two or more stress fields of different origin are superimposed to give a combined stress field. An example of such a combined field is shown in Figure 9.9. Stresses at any point may be combined by calculating each set of stresses in the form of stress components with reference to the same set of axes x, y and z. The combined stress system is found by adding the components, e.g. $\sigma_x = \sigma_{x_1} + \sigma_{x_2}$, $\tau_{xy} = \tau_{x_1 y_1} + \tau_{x_2 y_2}$, etc. The new principal stresses may then be found by calculating positions for which $\tau = 0$. The method for calculating the principal stress axes given six stress components is given by Ramsay (1967), pp. 31–34.

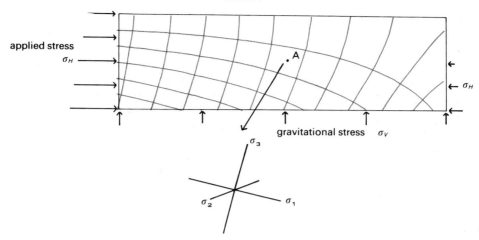

Figure 5.11 Stress trajectories. The diagram shows theoretical stress trajectories (colour) in a rectangular block of crust subjected to a variable horizontal stress σ_H applied to the sides of the block and a uniform vertical gravitational stress σ_V. The intermediate principal stress σ_2 is perpendicular to the plane of the diagram. The stress axial cross at any point A can be found by interpolation. After Hafner, W. (1951) *Bull. geol. Soc. Am.* **62**, 373–398.

Further reading

Jaeger, J.C. and Cook, G.G.W. (1976) *Fundamentals of Rock Mechanics*, Chapman and Hall, New York. Gives a comprehensive treatment of the physics of stress and the behaviour of materials.

Means, W.D. (1976) *Stress and Strain*, Springer Verlag, New York. Gives a thorough account of stress as an aspect of continuum mechanics, but in a reasonably elementary way aimed specifically at the geologist. Easier to follow than Jaeger and Cook.

Ramsay, J.G. (1967) *Folding and Fracturing of Rocks*, McGraw-Hill, New York.

6 Strain

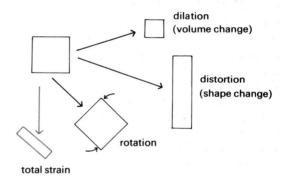

Figure 6.1 The nature of strain: dilation, distortion and rotation.

6.1 Nature of strain

As explained in the previous section, strain is the geometrical expression of the amount of deformation caused by the action of a system of stresses on a body. We can thus define strain as the change in size and shape of a body resulting from the action of an applied stress field.

Strain is expressed as *dilation* (volume change) or *distortion* (shape change), or as a combination of these processes. In addition it is often convenient to describe the distortion of a body in terms of a non-rotational or *coaxial* shape change plus a *rotational* component (Figure 6.1).

Homogeneous and inhomogeneous strain. If the amount of strain in all parts of a body is equal, the strain is said to be *homogeneous* (Figure 6.2A). The criteria for *homogeneous strain* are that straight lines remain straight and that parallel lines remain parallel. In the case of *heterogeneous* (inhomogeneous) strain, the strain in different parts of a body is unequal (Figure 6.2B). The criteria for *heterogeneous strain* are thus that straight lines become curved and that parallel lines become non-parallel.

The difference between homogeneous and

A homogeneous strain

B inhomogeneous strain

Figure 6.2 Homogeneous (*A*) and inhomogeneous (*B*) strain (see text).

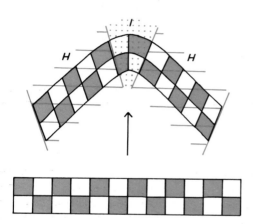

Figure 6.3 Domains of homogeneous (*H*) and heterogeneous (*I*) strain in a folded layer (see text).

heterogeneous strain can be simply illustrated by the folded layer of Figure 6.3. Taken as a whole, the fold exhibits heterogeneous strain. However, the straight limbs of the fold taken separately exhibit homogeneous strain. This is an example of a very useful principle in strain analysis, which is that complex heterogeneous strains are most conveniently studied by breaking them down into smaller homogeneous domains (cf. the study of fabric in section 3.4, Figure 3.9).

6.2 Measurement of strain

Strain may be measured in two ways (Figure 6.4): either by a change in length of a line or by a change in the angle between two lines. Any strain geometry can be measured as a combination of these changes. They are defined thus:

(i) *extension*:

$$e = (l - l_0)l_0 \tag{6.1}$$

where l_0 is the original length and l the new length of a line. Note that a positive value of e is an *elongation* whereas a negative value of e is a *shortening*.

Alternatively, the change in length of a line may be given by the *stretch*, defined as the ratio of the new length to the old length. Thus:

$$e = l/l_0 \tag{6.2}$$

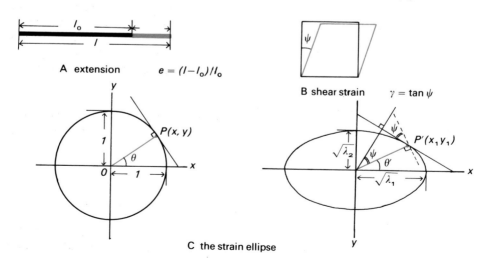

A extension $e = (l-l_0)/l_0$

B shear strain $\gamma = \tan \psi$

C the strain ellipse

Figure 6.4 Extension, shear strain and the strain ellipse. *A*. Extension $e = (l - l_0)/l_0$. *B*. Shear strain $\gamma = \tan \psi$. *C*. The strain ellipse (see text for explanation).

For many purposes this measure of strain is more convenient. In the study of large crustal deformations, the stretch is usually termed the β (beta) *factor*.

Another expression of the change in length of a line is the *quadratic elongation* λ (lambda) given by

$$\lambda = (l/l_0)^2 = (1 + e)^2 \qquad (6.3)$$

(ii) *shear strain*:

$$\gamma\,(\text{gamma}) = \tan \psi\,(\text{psi}) \qquad (6.4)$$

where ψ is the deflection of an originally right angle.

e, γ, λ and ψ are all dimensional quantities measuring the strain in a particular direction.

Strain in two dimensions. Consider a circle of unit radius deformed into an ellipse with major axis $\sqrt{\lambda_1}$ and minor axis $\sqrt{\lambda_2}$ (Figure 6.4C). This ellipse is known as the *strain ellipse*. The point $P(x, y)$ on the unit circle is transferred to $P'(x_1, y_1)$ on the ellipse. If θ is the angle made by OP and the x axis before deformation, and θ' the angle after deformation, then, since $x_1 = x\sqrt{\lambda_1}$ and $y_1 = y\sqrt{\lambda_2}$,

$$\tan \theta' = \frac{y_1}{x_1} = \frac{y\sqrt{\lambda_2}}{x\sqrt{\lambda_1}} = \tan\theta\,\frac{\sqrt{\lambda_2}}{\sqrt{\lambda_1}}$$

therefore $\qquad\qquad\qquad\qquad (6.5)$

$$\tan\theta'/\tan\theta = \sqrt{\lambda_2}/\sqrt{\lambda_1}$$

The length of the line OP is changed by an amount $(1 + e)$, such that

$$(1 + e)^2 = \lambda = x_1^2 + y_1^2$$
$$= \lambda_1 \cos^2 \theta + \lambda_2 \sin^2 \theta \qquad (6.6)$$

The relationship between shear strain and extension is given by

$$\gamma = \tan\psi = \left(\frac{\lambda_1}{\lambda_2} - \frac{\lambda_2}{\lambda_1} - 2\right)^{1/2} \cos\theta\sin\theta \qquad (6.7)$$

See Ramsay (1967) 67–68 for proof.

Strain in three dimensions. The strain of a body can be measured in three dimensions with reference to three arbitrarily chosen coordinate

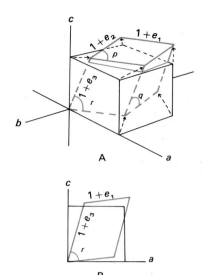

Figure 6.5 Strain components in three dimensions: the strain parallelepiped (see text for explanation). *A*, in three dimensions; *B*, in the plane *ac*.

axes a, b and c in the same way as for stress (see section 5.5). By taking a very small (infinitesimal) cube with sides parallel to a, b and c, we can describe the strain 'at a point' with reference to the change in shape of the cube, which becomes a parallelepiped—the *strain parallelepiped* (Figure 6.5). The *infinitesimal strain components* are the extensions and deflections with reference to axes a, b and c necessary to transform the cube into the strain parallelepiped. If the original cube has sides of unit length, then the sides of the parallelepiped are $(1 + e_1)$, $(1 + e_2)$ and $(1 + e_3)$, where e_1 is the extension in $a, \ldots$, etc. (or θ_1, θ_2 and θ_3, where θ_1 is the stretch in $a, \ldots$, etc.) and the angles of the parallelepiped are p, q and r such that:

$$p = 90° - \tan^{-1} \gamma_{ca}$$
$$q = 90° - \tan^{-1} \gamma_{ab} \qquad (6.8)$$
$$r = 90° - \tan^{-1} \gamma_{bc}$$

where γ_{ca} is the shear strain of c with respect to $a, \ldots$, etc. The six independent strain components are thus e_1, e_2 and e_3; γ_{ab}, γ_{bc} and γ_{ca}. These provide a complete description of the strain.

6.3 Principal strain axes and the strain ellipsoid

An alternative and more useful way to describe the strain is to select three mutually perpendicular axes x, y and z such that they are parallel respectively to the directions of greatest, intermediate and least elongation of the strain parallelepiped. The axes x, y and z are known as the *principal strain axes*. In this form the strain is more conveniently regarded as an ellipsoid, the *strain ellipsoid* (Figure 6.6), with maximum, intermediate and minimum axes X, Y and Z, representing the stretches along x, y and z. This ellipsoid is the shape taken up by a deformed infinitesimal sphere of unit radius. X, Y and Z are known as the *principal strains*. In terms of extensions, $X = 1 + e_x$, $Y = 1 + e_y$ and $Z = 1 + e_z$, where e_x, e_y and e_z are the extensions along x, y and z respectively. To complete the description of the geometry of the strain, the orientations of X, Y and Z with respect to reference axes a, b and c have to be given in addition.

6.4 Pure shear and simple shear (distortion and rotation)

If the orientations of the principal strains X, Y and Z have not changed during the deformation, the strain may be described as *irrotational* or *coaxial*, and this process is generally known as *pure shear* (Figure 6.7A). Where a change in orientation has occurred, the strain is described as *rotational* (*non-coaxial*) and this process is known as *simple shear* (Figure 6.7B). The difference is more easily portrayed in two dimensions.

A strain may thus be described in terms of a *distortional component*, which measures the ellipsoid shape, plus a *rotational component*, which measures the rotation of the principal strain axes from their original attitudes in the unstrained state.

6.5 Special types of homogeneous strain

It is convenient to recognize three special cases of homogeneous strain which can be distinguished by particular ratios of the principal strains X, Y and Z. In the general case, the three are unequal

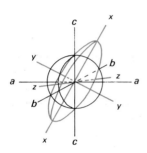

Figure 6.6 The strain ellipsoid: principal strain axes x, y and z (see text for explanation).

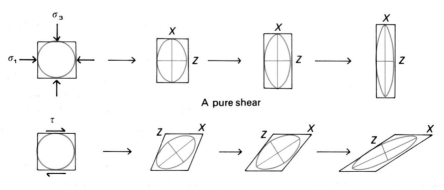

A pure shear

B simple shear

Figure 6.7 Pure shear and simple shear. In the process of pure shear (*A*), which involves irrotational or coaxial strain, the orientations of the principal strains X and Z do not change during progressive deformation. In simple shear (*B*), which involves rotational or non-coaxial strain, principal strains X and Z rotate in a clockwise manner during progressive deformation.

and $X > Y > Z$. The special cases, shown in Figure 6.8 are:

(i) *Axially symmetric extension* $(X > Y = Z)$—Figure 6.8*A*. This type of strain involves uniform extension in the X direction and equal shortening in all directions at right angles to it. This corresponds to a *prolate* or 'rugby ball' type of strain ellipsoid.

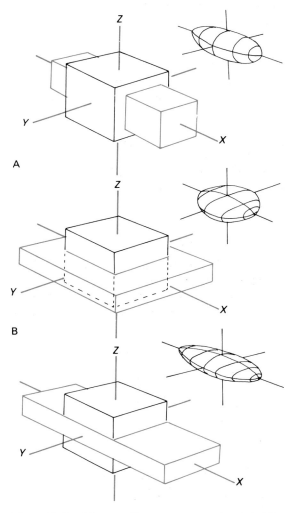

A

B

Figure 6.8 Special types of homogeneous strain. *A*. Axially symmetric extension $(X > Y = Z)$. This is a prolate uniaxial ellipsoid. *B*. Axially symmetric shortening $(X = Y > Z)$. This is an oblate uniaxial ellipsoid. *C*. Plane strain $(X > Y = 1 > Z)$. This is a triaxial ellipsoid where the intermediate axis is unchanged.

(ii) *Axially symmetric shortening* $(X = Y > Z)$—Figure 6.8*B*. This type of strain involves uniform shortening in the Z direction and equal extension in all directions at right angles to it. This corresponds to an *oblate* or 'pancake' type of strain ellipsoid.

(iii) *Plane strain* $(X > Y = 1 > Z)$—Figure 6.8*C*. This type of strain is distinguished by the intermediate principal stress axis remaining unchanged (i.e. Y has unit length). X is extended and Z is shortened. Thus plane strain is a special type of triaxial ellipsoid.

6.6 Volume change during deformation

Changes in volume commonly accompany shape changes during deformation, and if these are not recognised they can cause misleading estimates of the principal strain ratios. The volume change or *dilation* Δ (delta) is given by

$$\Delta = (V - V_0)/V_0 \qquad (6.9)$$

where V and V_0 are the volumes in the deformed and undeformed states respectively.

Since the volume of the strain ellipsoid is $\frac{4}{3}\pi$ $(X \times Y \times Z)$ derived from a unit sphere of volume $\frac{4}{3}\pi$ the dilation is given by

$$\Delta = (X \times Y \times Z) - 1 \qquad (6.10)$$

or as

$$1 + \Delta = (1 + e_x)(1 + e_y)(1 + e_z) \qquad (6.11)$$

where e is the extension.

6.7 Graphical representation of homogeneous strain

A convenient way of expressing the various strain states is to use the *Flinn diagram* (Flinn, 1962). In this diagram (Figure 6.9*A*), the ratios of the principal strains are taken, such that

$$a = X/Y = (1 + e_x)(1 + e_y) \qquad (6.12)$$

and

$$b = Y/Z = (1 + e_y)(1 + e_z) \qquad (6.13)$$

and a is plotted against b.

The different shapes of ellipsoids are distinguished using the value k, such that

$$k = (a - 1)/(b - 1) \qquad (6.14)$$

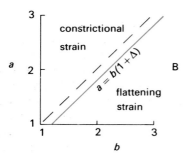

Figure 6.9 Graphical representation of strain ellipsoids: the Flinn diagram. A. Different ellipsoids are described using the value $k = (a-1)/(b-1)$—see text for explanation. B. If the volume is not constant, the line $a = b(1 + \Delta)$ divides the field of constrictional strain from the field of flattening strain. The diagram shows the effect of a 20% volume reduction. After Flinn, D. (1962) *Quart. J. geol. Soc. Lond.* **118**, 385–433, Ramsay and Huber (1983).

The various strain states can be described as follows:

Axially symmetric extension: $k = \infty$
Constrictional strain (prolate ellipsoids): $1 < k < \infty$
Plane strain (at constant volume): $k = 1$
Flattening strain (oblate ellipsoids): $0 < k < 1$
Axially symmetric flattening: $k = 0$

In this way the shape of an ellipsoid can be described using only the value of parameter k, and constrictional and flattening strains can immediately be distinguished by whether k is greater or less than 1.

Figure 6.9A is constructed assuming a constant volume, since the line $k = 1$ will only pass through the origin when the volume change $\Delta = 0$. When $\Delta \neq 0$, equation (6.9) reduces to

$$1 + \Delta = (1 + e_x)(1 + e_z)$$

(since $(1 + e_y) = 1$ for $k = 1$ ellipsoids)

$$= a/b$$

therefore

$$a = b(1 + \Delta) \qquad (6.15)$$

Thus for a volume change of Δ, the line $a = b(1 + \Delta)$ represents plane strain and divides the constrictional from the flattening fields (Figure 6.9B).

6.8 Progressive deformation and finite strain

The strained body at the time of measurement represents the total strain produced in the body up to that time, and is produced by adding a series of *strain increments* as the body takes up a succession of different shapes and positions in response to the applied stress (Figure 6.10). This process, from initial to final position, is termed *progressive deformation* and the final strain at the time of measurement is termed the *finite strain*. It is important to recognize that the nature of the finite strain is not necessarily a reliable guide to the various intermediate states of strain. Even quite simple cases of progressive strain display marked changes in strain pattern with time, and may involve, for example, sets of material lines that pass through a stage of contraction followed by extension as deformation progresses.

At any given instant during progressive deformation, it is possible to examine both the *finite strain* (total strain up to that time) and the *infinitesimal strain* at that point in time. In two dimensions, the finite-strain ellipse (Figure 6.11A) can be divided into sectors of elongation and contraction separated by lines of no finite longitudinal strain (i.e. zero extension). Suitably oriented layers will exhibit *boudinage* in elongation sectors and *folding* in contraction

finite strain

unstrained

Figure 6.10 Progressive deformation. The finite strain is achieved by adding successive strain increments to the initial unstrained shape.

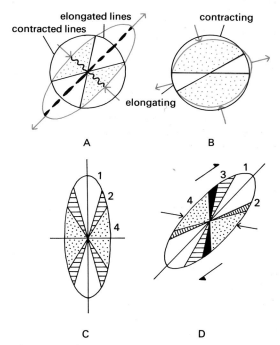

Figure 6.11 Changing fields of elongation and contraction during progressive deformation (in two dimensions). *A.* The finite strain ellipse. Fields of elongated lines (boudinage) and contracted lines (folds) are separated by lines of zero extension which retain their original length (= radius of undeformed circle). *B.* The infinitesimal strain ellipse. Fields that at any instant in time during progressive deformation are currently contracting or elongating. *C.* Superimposition of ellipses *A* and *B* for pure shear will produce three zones: 1, continued elongation; 2, elongation followed by contraction; and 4, continued contraction. *D.* Superimposition of ellipses *A* and *B* for simple shear will produce an asymmetric arrangement of four zones: 1, continued elongation; 2, elongation followed by contraction; 3, contraction followed by elongation; and 4, continued contraction. After Ramsay (1967), Figs. 3.56 and 3.62.

sectors. The infinitesimal strain ellipse (Figure 6.11*B*) at that instant will show sectors whose lines are currently elongating or contracting.

By superimposing the two ellipses, we can distinguish four zones:

zone 1: continued elongation (boudins)
zone 2: elongation followed by contraction (folded boudins)

zone 3: contraction followed by elongation (unfolded or boudinaged folds)
zone 4: continued contraction (folds).

The distribution of these zones or sectors will depend on the strain history and, in particular, on whether the strain is *irrotational* (i.e. *pure shear*—Figure 6.11*C*) or *rotational* (i.e. *simple shear*—Figure 6.11*D*). Observations of the orientations of folded and boudinaged layers are therefore of great value in investigating progressive deformation.

Growth fibre analysis. Another very useful method of investigating the strain history is to analyse the orientation of quartz or calcite growth fibres commonly found in extensional cracks or veins. These growth fibres are crystals oriented parallel to the direction in which the walls of the crack have opened and thus to the direction of extension in the rock. Changes in fibre orientation mark changes in the extension direction. Often quite complex changes in direction are recorded in this way. The process of periodic opening of cracks and filling with vein material is known as the *crack-seal mechanism* and is described in detail by Ramsay and Huber (1983).

6.9 Relationship between stress and strain

Since a strain results from the action of a stress, there must always be a direct geometrical relationship between the two. However, since the geometry of both stress and strain fields changes with time, the relationship may not be a simple one. We can illustrate this principle with reference to the two cases of homogeneous strain shown in Figure 6.7—*pure shear* and *simple shear*.

In the case of *pure shear* (Figure 6.7*A*), the orientation of the principal finite strain axes and the principal stress axes correspond, with $X \parallel \sigma_3$, $Y \parallel \sigma_2$ and $Z \parallel \sigma_1$; that is, the direction of greatest extension corresponds to the direction of minimum stress (which in many cases would be negative, or tensional) and the direction of least extension or greatest shortening corresponds to the direction of maximum stress.

However, in the simple shear case (Figure 6.7B), only the intermediate stress and strain axes correspond (since it is plane strain), and the X and Z axes rotate progressively clockwise with increasing strain away from their initial position; they will not in general correspond with the orientations of the σ_1 and σ_3 stress axes.

Since only the strain axes are observed in natural deformation, the position of the stress axes cannot immediately be deduced. It can be seen from Figure 6.7 that the magnitudes of the strain axes at stages 2, 3 and 4 of A are respectively identical with those of B although they have been produced differently. The change in shape of a strained body will itself cause a change in the stress field. This is perhaps more obvious in the case of simple shear. When the sides of the strained body of Figure 6.7B depart from their initial perpendicular relationship, the normal and shear stresses on them will be orientated obliquely to the original orthogonal set. Thus the position of the stress axes can only be reconstructed by reference to the *strain history* of the deformed body.

Examples of how the stress field may be reconstructed from fault displacement data are discussed in section 9.2.

Further reading

Flinn, D. (1962) On folding during three-dimensional progressive deformation. *Quart. J. geol. Soc. Lond.* **118**, 385–433.

Means, W.D. (1976) *Stress and Strain*, Springer Verlag, New York.

Ramsay, J.G. (1967) *Folding and Fracturing of Rocks*, McGraw-Hill, New York.

Ramsay, J.G. and Huber, M.I. (1983) *The Techniques of Modern Structural Geology*, Vol. 1: *Strain Analysis*, Academic Press, New York.

Skjernaa, L. (1980) Rotation and deformation of randomly oriented planar and linear structures in progressive simple shear. *J. struct. Geol.* **2**, 101–109.

7 Stress and strain in materials

The way in which individual minerals respond to stress varies widely according to the physical conditions under which the deformation takes place, and depends also on the compositional and mechanical properties of the material. Before considering the behaviour of rocks under stress, we shall discuss various 'ideal' types of response.

7.1 Ideal elastic and viscous strain

Elastic strain. In ideal elastic strain, the removal of the deforming stress causes an immediate return of the body to its original unstrained shape. This type of strain is often called *temporary* or *recoverable* strain and can be demonstrated by the compression and release of a spring. It corresponds to the type of strain associated for example with the propagation of seismic waves through the Earth, or indeed with the passage of sound waves through any medium.

The behaviour of perfectly elastic bodies is governed by *Hooke's Law*, which states that

$$e = \sigma/E \qquad (7.1)$$

i.e. stress/strain = a constant

where e is the extensional strain, $= (l - l_0)$, where l is the new length, l_0 the original length, σ the applied stress and E a constant known as *Young's Modulus* or the *elasticity* of the material. Also

$$e_V = P/K \qquad (7.2)$$

where e_V is the dilational strain, $= (V - V_0)/V_0$, where V is the new volume, V_0 the original volume, P the hydrostatic pressure, and K a constant termed the *compressibility*. Thus in elastic shape and volume changes, the strain is directly proportional to the stress, that is, stress and strain have a linear relationship (Figure 7.1A). Rocks exhibit perfectly elastic behaviour only under certain restricted conditions.

Viscous strain. In ideal viscous strain, there is no recovery after removal of the deforming stress; that is, all the movement is permanent. Ideal viscous or 'Newtonian' behaviour is exhibited by the flow of fluids and is governed by the equation

$$\sigma = \eta \dot{e} \qquad (7.3)$$

i.e.

stress = viscosity × strain-rate

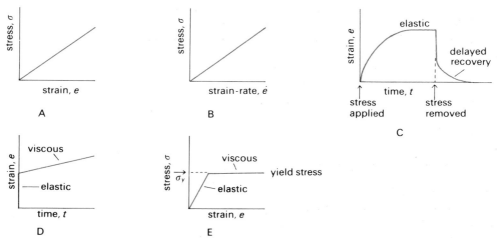

Figure 7.1 Deformation of material: types of ideal strain. *A.* Ideal elastic strain: $\sigma \propto e$. *B.* Ideal viscous strain: $\sigma \propto \dot{e}$. *C.* Strain-time diagram for ideal viscoelastic strain. The strain is temporary (recoverable) but is not instantaneous as in ideal elastic strain. *D.* Ideal elastoviscous strain at constant stress—an instantaneous elastic strain is followed by viscous strain for a stress of longer duration. *E.* Ideal plastic strain—elastic strain at low values of stress is replaced by viscous strain above the yield stress.

where $\dot{e}$ is the strain-rate (rate of change of shape with time) and η (eta) is a constant termed the *viscosity* of the material (for more realistic equations of flow the 'effective viscosity' varies with both temperature and stress). Thus in ideal viscous strain, the stress is linearly related to the strain rate (Figure 7.1*B*) so that the higher the applied stress, the faster the material will deform, and the total strain is dependent both on the magnitude of the stress and the length of time for which it is applied.

For a constant stress, the strain will increase linearly with time t, since, integrating equation (7.3),

$$e = \sigma t/\eta \qquad (7.4)$$

7.2 Viscoelastic, elastoviscous and plastic behaviour

Real rock materials combine the properties of ideal elastic and viscous bodies, and the strain in such materials may be regarded as having both elastic and viscous components. One approximation to the total strain may be obtained by adding equations (7.1) and (7.4) to give

$$e = \sigma/E \quad \text{(elastic component)}$$
$$+ \sigma t/\eta \quad \text{(viscous component)} \qquad (7.5)$$

at *constant stress*. However, this equation oversimplifies the relationship and several other factors have to be taken into account.

Viscoelastic behaviour. A material that, for a given stress, exhibits a basically elastic type of strain, but which takes a certain time to reach its limiting value, is said to show *viscoelastic* behaviour (Figure 7.1*C*). Conversely, the removal of the stress does not cause an immediate return to the undeformed state, but there is a *delayed recovery* of the elastic strain. Delayed recovery is responsible for the well-known phenomenon of *earthquake aftershocks* which represent continued movements following the main release of strain represented by the earthquake (see section 9.3). Most rocks exhibit viscoelastic behaviour at low values of stress.

Elastoviscous behaviour. A material that basically obeys the viscous law (equation 7.7), but which behaves elastically for stresses of short duration, is termed *elastoviscous* (Figure 7.1*D*). Pitch is a good example of such a material. When stressed, it will show elastic strain which is completely recoverable if the stress is rapidly removed. The material will flow, however,

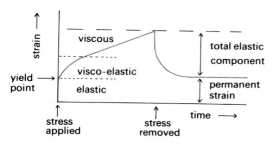

Figure 7.2 Ideal strain–time relationship for a typical plastic material deformed above its yield point.

exhibiting perfectly viscous behaviour, for a stress held for any length of time.

Plastic behaviour. A *plastic* material is one that behaves elastically at low values of stress but above a certain critical value of stress (the *yield stress*—see section 7.4) it behaves in a perfectly viscous manner (Figure 7.1E).

The behaviour of rock materials under stress cannot be described in terms of any one of these simple models but includes their characteristics in combination. Figure 7.2 shows diagrammatically how the strain–time diagram of a typical plastic material (like most rocks) exhibits elements of all the above models. The term *viscoelastic* is also used in a rather looser sense to describe any combination of elastic and viscous behaviour. In practice, the relationship between stress and strain in rock deformation varies both with the magnitude of the stress and with the strain rate.

7.3 Brittle and ductile behaviour

Where elastic deformation leads to *failure*, the material loses cohesion by the development of a fracture or fractures across which the continuity of the material is broken. This type of behaviour is called *brittle* behaviour and governs the development of faults and joints (section 9). *Ductile* behaviour, in contrast, produces permanent strain that exhibits smooth variations across the deformed sample or rock without any marked discontinuities. Most rock materials are capable of exhibiting either brittle or ductile behaviour depending on such factors as the size of the

differential stress ($\sigma_1 - \sigma_3$), the hydrostatic pressure, the temperature, the strain rate and the fluid pressure.

7.4 The effects of variation in stress

In this section we consider the effects of the *differential stress* ($\sigma_1 - \sigma_3$). The hydrostatic part of the stress, or *confining pressure*, is considered separately. Figure 7.3 summarizes diagrammatically the effects of increasing differential stress on the strain–time curve. For a low value of stress σ_A the material may exhibit entirely elastic behaviour. For slightly larger values of stress σ_B the strain may be partly viscoelastic but there is a limiting value of strain which is basically elastic. Above a certain critical value of stress known as the *yield stress* σ_Y the material exhibits essentially viscous behaviour for successively higher values of stress σ_C, σ_D after an initial viscoelastic strain. Above a second critical value of stress known as the *failure stress* (σ_R) the material exhibits accelerated viscous flow for higher values of stress leading to failure after the initial viscoelastic and viscous stages (compare the creep curve, Figure 7.4).

We may distinguish three main fields on a typical strain–time diagram corresponding to progressive increase in stress—*elastic*, *viscous* and *failure*. In the case of a *ductile* substance the viscous field is enlarged at the expense of the elastic and failure fields, and $\sigma_R \gg \sigma_Y$.

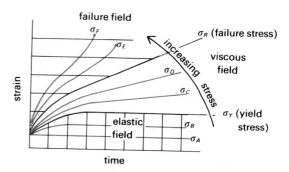

Figure 7.3 Diagrammatic representation of the effect of increasing stress on a strain–time curve (see text for explanation).

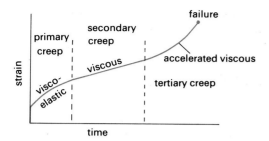

Figure 7.4 Creep: strain–time diagram for long time periods.

Strength of materials. The measured strength of a material is simply the value of the applied stress at which failure occurs. Many materials possess both a *yield strength* (σ_Y) defined as the limiting stress above which permanent deformation occurs, and a *failure strength* or *ultimate strength* (σ_R) above which failure occurs. The values of compressive and tensile strength are normally different (section 9.1)—i.e. the yield strength and failure strength are usually higher under a compressive differential stress ($\sigma_1 - \sigma_3$ positive) than under a tensile differential stress ($\sigma_1 - \sigma_3$ negative).

Because of the effect of creep (see below), values of strength measured over short time periods ('*instantaneous*' strength) are much larger than those measured (or extrapolated) for long periods of time. The long term or *creep strength* of most rocks is only in the range 20–60% of their instantaneous strength (see effect of strain rate, section 7.9).

7.5 Interrelationship of stress, strain and time

The relationship between stress and strain for real materials, which exhibit a combination of elastic, viscous and plastic properties, depends critically on the length of time for which the stress is applied. In laboratory experiments on rocks, with short durations of seconds or minutes, the behaviour of the material is effectively 'instantaneous' in geological terms and differs significantly from that of the same material under stresses with more geologically realistic durations of months or years. The long-term strain behaviour of materials is termed *creep*. The

important characteristic of creep behaviour is that viscous strain is produced over long periods of time under low stresses that would produce only elastic effects if applied over short periods. Another way of expressing this difference in behaviour is to state that the *strain rate* has a decisive influence on the type of strain behaviour: that is, materials deforming under low strain rates that are geologically realistic exhibit creep, whereas high strain rates (e.g. of experimental deformation) are associated with geologically instantaneous strain.

The typical creep curve. Most rocks when subjected to a constant, low stress over a period of, say, months exhibit a strain curve that consists of three stages (Figure 7.4): an initial *primary* or *transient creep* where the material behaves visco-elastically, followed by a stage of *secondary* or *steady-state creep* where the material exhibits essentially viscous flow, and a final stage of *tertiary creep* where the material exhibits accelerated viscous strain leading ultimately to failure. Creep at constant stress may be represented by an equation of the form

$$e = A + B \log t + Ct \qquad (7.6)$$

where A, B and C are constants reflecting the physical properties of the material and the values of the stress and temperature (cf. equation 7.5). Note the resemblance of the typical creep curve to the 'ideal' strain–time curve of Figure 7.2.

7.6 The effect of confining pressure

In considering the effect of an applied stress field during deformation, the hydrostatic and deviatoric stress components must be considered separately (see section 5.6). Rocks at depth in the crust are subjected to the load or *lithostatic* pressure of the overlying column of rock. This pressure can be assumed to be effectively hydrostatic, and is simply related to the thickness and mean density of the overlying material. The pressure at the base of a 35-km thick crust is about 10 kilobars, and realistic pressures for most naturally deformed crustal rocks vary from several hundred bars upwards.

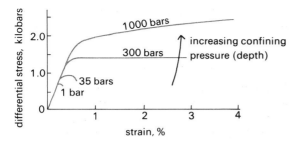

Figure 7.5 Effect of increasing confining pressure on the stress–strain curves for the experimental deformation of Wombeyan marble. Increase in confining pressure raises the yield stress or strength of the rock. After Paterson, M.S. (1958). *Bull. geol. Soc. Am.* **69**, 465–476.

The hydrostatic stress (often called *confining pressure*) causes elastic volume changes that depend on the compressibility of the material (see equation 7.2). The size of these volume changes is unimportant except at great depth. Primary seismic waves propagate by such elastic volume changes by means of alternate compression and dilation of the material through which the waves are transmitted.

A more important aspect of confining pressure in the study of strain is its effect on the *strength* of materials. With increasing confining pressure, both the yield stress σ_Y and the failure stress σ_R are raised, giving the material a higher effective strength. Figure 7.5 illustrates the effect of confining pressure on the stress–strain curves for the experimental deformation of marble. At low pressures the response is basically elastic and failure occurs at low values of σ. At 300 bars confining pressure, the yield stress is raised to around 1400 bars and is followed by a steady viscous increase in strain, indicating that the material has become ductile at high pressure.

7.7 The effect of temperature

With increasing temperature, the yield stress σ_Y is lowered and the failure stress σ_R is raised, which has the effect of enlarging the viscous field of deformation at the expense of the elastic and failure fields (cf. Figure 7.3). Consequently the material may be said to show an increase in ductility.

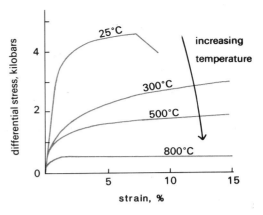

Figure 7.6 Effect of increasing temperature on the stress–strain curves for the experimental deformation of Yule marble at 5 kbars confining pressure. Increase in temperature decreases the yield stress or strength of the rock. After Griggs, D.T., Turner, F.J. and Heard, H.C. (1960) in *Rock Deformation*, eds. D.T. Griggs and J. Handin, *Mem. geol. Soc. Am.* **79**.

The effect of temperature on the stress–strain curves for the experimental deformation of marble is shown in Figure 7.6. The yield stress at 800 °C is less than 0.5 kilobars, about one-sixth of its value at room temperature. Moreover the strain range of viscous behaviour is considerably increased.

These observations are consistent with our geological experience of metamorphic rocks deformed at elevated temperature and pressure. These exhibit much more ductile types of deformation than do the equivalent rocks at the surface.

7.8 The effect of pore-fluid pressure

The presence of a fluid phase in rocks undergoing deformation is important in several ways. The most important effect is to facilitate deformation by reducing resistance to slip along planes of potential movement within a rock, from grain boundaries to major fault planes (see also section 9.1). The fluid pressure has the effect of reducing the shear stress required for slip; i.e. it reduces the shear strength of a rock. This is because the direct pressure between adjoining grains caused by the lithostatic or confining pressure is countered by the effect of the pore-

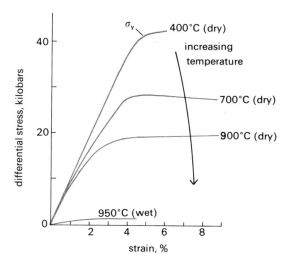

Figure 7.7 Effect of pore-fluid pressure. Stress–strain curves for the deformation of wet and dry natural quartz crystals at 15 kbars confining pressure, various temperatures and a strain rate of 0.8×10^{-5}/s. See text for explanation. After Griggs, D.T. (1967) *Geophys. J. Roy. Astr. Soc.* **14**, 19–31.

fluid pressure. Fluid pressure may also promote mineralogical reactions, particularly at elevated temperatures, that reduce the mechanical strength of the rock.

The mechanical effect of the pore-fluid pressure is given by

$$P_e = P - P_f$$

where P_e is the *effective pressure* on the solid material, P is the confining pressure, and P_f the fluid pressure.

For saturated rocks, where the pore-fluid pressure may be very high compared with the confining pressure, the effect of the high confining pressure is cancelled out and the strength of a rock reduced to near-surface conditions. The effect of high fluid pressure on rock at elevated temperature is illustrated by the stress-strain curves for wet and dry quartz crystals (Figure 7.7). The yield stress at 950 °C in wet quartz is only about one-tenth of that required for the dry material at the same temperature. Clearly the ductility in this case is greatly increased, which explains why certain materials, normally strong even at high temperature, can

flow under metamorphic conditions in the presence of aqueous fluids.

7.9 The effect of strain-rate

The importance of strain-rate in rock deformation has already been referred to in the discussion of creep. Values of the yield and ultimate strengths of rocks are much higher if measured over short time periods than over geologically significant time periods. Deformation experi-

A

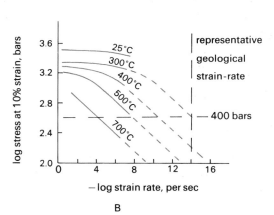

B

Figure 7.8 Effect of strain-rate on the stress–strain relationship. *A.* Stress–strain curves for Yule marble deformed by extension at 600 °C at strain rates varying from 2×10^{-3} s^{-1} to 2×10^{-7} s^{-1} (*a*, 2×10^{-3}; *b*, 2×10^{-4}; *c*, 2×10^{-5}; *d*, 2×10^{-6}; *e*, 2×10^{-7} s^{-1}). *B.* Stress–strain curves for deformation of Yule marble at various temperatures, plotted as log stress at 10% strain (approx. = yield stress) against log strain rate. *A* and *B* after Heard, H.C. and Raleigh, C.B. (1972) *Bull. geol. Soc. Am.* **83**, 935–956. The curves representing the experimental data are extrapolated (dashed lines) to show the effect of lower strain rates (see text for further explanation).

ments lasting for days rather than seconds have been designed to investigate the effect of lower strain rates. A succession of stress–strain curves for Yule marble (Figure 7.8A) at strain rates varying from $2 \times 10^{-3} \, s^{-1}$ ($= 20\%$ in 100 s) to $2 \times 10^{-7} \, s^{-1}$ ($= 20\%$ in 12 days) demonstrate a gradual reduction of strength and increase in ductility with decrease in strain rate.

A strain rate of $2 \times 10^{-7} \, s^{-1}$ is still much greater than geological strain rates which are around $10^{-14} \, s^{-1}$—i.e. 10% in one million years. In order to extrapolate the experimental results to such a low value, a log stress/log strain rate plot has been used (Figure 7.8B). This plot indicates that at a temperature of 300 °C the differential stress required to maintain steady ductile flow at a strain rate of $10^{-14} \, s^{-1}$ is only about 400 bars (40 MPa), or 160 bars (16 MPa) at 400 °C.

7.10 Summary: physical controls on strain behaviour

Brittle failure is typical of rocks at low confining pressure and low temperature appropriate to near-surface conditions. In the temperature–pressure range found throughout the greater part of the crust (e.g. confining pressures of 0.1–3 kilobars (10–300 MPa) and temperatures of 100–500 °C) most rocks show at least some ductile flow before failure. The yield strength, the critical value of stress difference required to initiate this ductile behaviour, is rather high in many rocks and would effectively inhibit yield under surface conditions. However, the yield strength is dramatically reduced by pore-fluid pressure (particularly at high temperatures) and, even more important, by decreasing the strain rate to geologically appropriate rates of about $10^{-14} \, s^{-1}$. Thus, given the physical conditions existing at some depth within the crust, and several million years under stress, most rocks will exhibit the kind of ductile behaviour familiar to all geologists who have studied folded rocks in metamorphic terrains. The same rocks under higher stresses and more rapid strain-rates, however, will fracture and generate earthquakes.

7.11 Mechanisms of rock deformation

Since rocks consist of aggregates of individual crystal grains, normally of several different mineral species, the way that they deform depends partly on the properties of the individual crystals and partly on the texture of the rock as a whole. An igneous rock with an interlocking crystalline texture will clearly be stronger than a sandstone with a weak carbonate cement, and stronger than a rock that is cut by pervasive planar fractures, regardless of the nature of the actual minerals. Much can be learned about the nature of rock deformation by studying the microscopic fabric of deformed rocks (see section 3.4) and it is possible in favourable cases to reconstruct in detail how the final strained shape of a deformed rock has been achieved by successive changes in crystal shapes and inter-relationships.

The purely elastic properties of a rock are conferred on it by elastic distortions of the lattice of individual crystals. When the lattice is subjected to a differential stress, the atomic spacing is slightly changed by an amount which is proportional to the size of the stress and which also depends on the interatomic bonding force— a characteristic property of the crystal, and hence of a particular rock type. This mechanism is responsible for the primary elastic stage of the typical creep strain curve.

Permanent strain is produced by various *deformation mechanisms* that operate on a microscopic scale. There are three main types of process: *cataclasis, intracrystalline plasticity* and *diffusive mass transfer*.

Cataclasis is the process of fracture and sliding of rigid particles. It includes *grain boundary sliding*, which is one of the most common mechanisms of deformation. It may be accompanied by diffusion, in which case it is called *superplasticity*, or by the transfer of material in solution, when it is termed *crystal plasticity*. Grain boundary sliding produces parallel alignment of grain boundaries and rectangular grain shapes. Individual particles are undistorted. This process is therefore characterized by a shape fabric, but not by a crystal orientation fabric.

Intracrystalline plasticity involves the movement of *dislocations* through the crystal lattice. This process results in a change of shape of the crystal, and a variety of characteristic crystal features form, including *undulose extinction, deformation bands, deformation twins, kink bands* and *deformation lamellae*. These features give rise to a *preferred orientation* of crystal domains, which may be enhanced by recrystallization. Thus this process, unlike grain boundary sliding, produces an *oriented crystal fabric*.

Diffusive mass transfer involves the movement of material at the molecular scale. Such movement may be in the solid state through the crystal lattice (*Coble creep*) or via grain boundaries (*Nabarro–Herring creep*); or it may occur in solution. The latter process, when controlled by stress, is termed *pressure solution*, and is important in the formation of structures such as certain *spaced cleavages* and *stylolites* (section 3.1).

Most rock deformation involves a combination of two or more of the above types of process. For example, diffusive mass transfer accompanied by grain boundary sliding (superplasticity) is thought to be responsible for the formation of very fine-grained *mylonites* (section 1.3). Since most rocks are polycrystalline aggregates, different crystal phases may exhibit different mechanisms; for instance feldspar may deform cataclastically whereas quartz exhibits intracrystalline plasticity under greenschist facies conditions. The most important factors controlling the deformation mechanism are (i) the physical properties of the material; (ii) the magnitude of the deviatoric and hydrostatic stress components; (iii) the pore-fluid pressure; (iv) the temperature; and (v) the grain size. All deformation processes may be enhanced by chemical reactions.

These processes are irreversible and under appropriate levels of stress and physical conditions give rise to uniform viscous flow—the secondary or steady-state stage of the creep strain curve of Figure 7.4. Accelerated viscous flow is usually caused by the spread of microfractures or slip surfaces through the rock in such a way that they link up to form continuous pervasive cracks causing loss of cohesion and failure.

The relative importance of slip-type displacements and recrystallization in promoting viscous flow is governed primarily by rock composition, temperature and the presence of pore-fluids. Low-temperature deformation involves mainly cataclasis and intracrystalline plasticity (or *cold-working*). Strain fabrics produced by both cataclasis and cold-working may be removed at elevated temperature by recrystallization. New unstrained and generally polygonal sub-grains develop from the original strained grains, which eventually become replaced. This process of strain recovery is known as *polygonization, hot-working* or *annealing*. The behaviour of crustal rocks below about 10–15 km (and also of upper mantle rocks) is largely governed by steady-state viscous flow at very low strain rates giving rise to a continuous and progressive cycle of distortion and annealing recrystallization. Under such conditions the strain-rate is governed only by the effective viscosity (cf. equation 7.3) of the material, and the size of the differential stress.

Further reading

Griggs, D.T. and Handin, J. (eds.) (1960) *Rock Deformation*. Mem. geol. Soc. Am. **79**.

Heard, M.C., Borg, I.Y., Carter, N.I. and Raleigh, C.B. (eds.) (1972) *Flow and Fracture of Rocks*. American Geophysical Union, Geophysical Monograph **16**.

Jaeger, J.C. and Cook, N.G.W. (1976) *Fundamentals of Rock Mechanics*, Chapman and Hall, New York.

8 Determination of strain in rocks

One of the most important objectives of the structural geologist is the quantitative evaluation of the total or bulk finite strain in a given area resulting from a deformation. Ideally, if we are to be able to evaluate, let us say, the effects of orogenic compression, we would wish to know the magnitudes and orientations of the three principal strain axes in all parts of the area. Once the strain distribution is known, in the form of a set of strain trajectories, or as a representative bulk strain ellipsoid, we can then try to explain it in terms of stress or movement models.

There are three quite different approaches to the problem of quantifying strain. The first and most obvious is to measure individual strain ellipsoids using various kinds of 'strain markers' and sum the results over the area in question. The second approach is to estimate the total shortening or elongation by examining the geometry of folds or faults (this method is, however, difficult to apply in three dimensions). The third, and in some ways the simplest, approach is to assume that the strain on a large scale is essentially homogeneous statistically, and that the statistical arrangement of all planar and linear structural elements throughout the area reflects both the orientation and size of the bulk finite principal strains. This approach has been particularly useful in dealing with highly-deformed zones in certain Precambrian gneiss terrains.

8.1 Finding the principal strain axes

The problem of determining the strain in a deformed rock is made much easier if the principal strain axes can be found first. This is possible if the deformed rock possesses a new planar or linear fabric that reflects the finite strain geometry. Thus a planar flattening fabric (slaty cleavage, schistosity or gneissosity) will lie in the XY plane of the strain ellipsoid (see sections 3.1 and 6.3) enabling the orientation of the Z strain axis to be found immediately. In certain circumstances, the orientation of X may also be found by observing the 'stretching direction' (i.e. an elongation lineation) in the plane of the foliation, and in this way all three strain axes may be uniquely determined. However, strongly linear fabrics ($X > Y \simeq Z$) or strongly planar fabrics ($X \simeq Y > Z$) will yield only one obvious principal strain axis.

Even if the orientation of all three principal strain axes can be measured, however, the complete determination of the strain requires the measurement of the extensions along each. This demands the use of some object or objects of known initial shape. Such objects are termed *strain markers*. We shall consider first the simplest type which initially had a spherical shape.

8.2 Use of initially spherical objects

A wide variety of strain markers has been used in strain studies. Much of the early work on strain determination was done on deformed fossils, but the simplest methods involve the measurement of initially spherical objects which assume an ellipsoidal form after deformation. The strain ellipsoid may be measured directly if the objects can be seen in three dimensions. Such objects include the ooids in oolitic limestones, the spherulites and vesicles found in volcanic rocks, reduction spots in slates, recrystallization spots in hornfels and various kinds of concretions. Since such objects are rarely perfectly spherical, some allowance has to be made for the initial variation in shape. This can be easily done if the shape variation is random by taking the geometric mean of a number of observations, but if there is some non-random variation in initial shape, controlled by bedding for instance, a more sophisticated technique must be employed (see later).

Another problem concerns the degree of homogeneity of strain through the rock. Many strain markers are made of material different from their matrix, so that the measured strain may not accurately reflect that of the whole rock. Thus hard siliceous or calcareous concretions would show much lower strains than a more ductile shaly matrix.

Direct measurement of strain axes. Any three non-parallel sections through the rock are sufficient to determine the strain axes, but the calculations are easier if the three sections are mutually perpendicular, especially if they are parallel to the principal strain planes. Measurements can be made directly from thin sections or polished slabs, or on enlarged photographs if the objects are small. To obtain an accurate result a number of measurements should be made. If these are plotted on a graph of long axis length against short axis length, the slope of the best-fit straight line through the origin gives the mean value of

Figure 8.1 Finding the principal strain ratio in two dimensions from deformed circular objects. The long axis and short axis of a number of ellipses (e.g. Fig. 8.2*A*) are plotted. The slope of the coloured line gives the ratio Y'/X'. After Ramsay and Huber (1983), Fig. 5.10*A*.

the strain ratio Y'/X' (ratio of minimum to maximum strain) in that plane (Figure 8.1). The visual method of plotting is preferable to simply calculating the arithmetic mean, since deviations from the straight line may be easily observed. Sometimes the larger particles give a different result from the smaller ones, due perhaps to differences in ductility.

Centre-to-centre method. If there is a large difference in ductility between the measured objects and their matrix, the measured strain will obviously not apply to the whole rock. In the case, say, of quartz pebbles in a mudstone matrix, much higher strains would take place in the matrix than in the pebbles. To deal with this problem, the centre-to-centre method may be used. This method relies on the fact that the distances between the centres of randomly arranged spheres are systematically altered during strain in such a way that the changes in distance are related to direction. The ratio between the minimum and maximum mean distances is equal to the strain ratio Y'/X'.

To determine this ratio, a plot is made of the distances between adjacent centres against the orientation of the line between the centres (Figure 8.2). The plot gives a minimum and maximum value (m_1 and m_2 respectively) for the distance, which is used to calculate the strain ratio, and two corresponding values for α (α_Y and α_X) for the orientation of Y' and X'.

Figure 8.2 Finding the principal strain ratio in two dimensions from the spacing of deformed objects. *A*. Distance between adjacent centres and angle α made with a reference line are measured for a number of adjacent centres. *B*. Plot of d against α yields a minimum value of d (m_1) and a maximum value (m_2) which represent respectively Y' and X'. The orientation of X' and Y' in relation to the reference line is given by α_x and α_y. After Ramsay and Huber (1983), Figs. 7.17, 7.18.

The above method has the advantage that it can be applied to any rock with randomly distributed particles regardless of their respective ductilities or whether they form perfect ellipses. The main problem is in determining accurately the positions of the centres of the particles.

8.3 Use of deformed conglomerates

Deformed conglomerates are much more common than other kinds of strain marker, and have therefore been used frequently in the field by structural geologists for assessing strain. Unfortunately they suffer from the serious disadvantage that their initial shape is generally not spherical. Moreover, they may possess a primary sedimentary orientation which, when added to the deformation fabric, produces a compromise fabric which cannot be simply analysed. Added to these complications is a further factor—the effect of ductility both between different types of pebble and between pebble and matrix. The centre-to-centre method described above would circumvent some of these problems, but centres of pebbles with a primary orientation fabric are unlikely to possess a random distribution—an essential precondition for this method. Much effort has been expended by structural geologists in devising satisfactory methods of analysing pebble strain. In general it is more useful to obtain a large number of approximate results from different localities and rock types than to concentrate effort in obtaining mathematically precise results from a single pebble bed that might be highly atypical.

If we can assume that the pebbles originally possessed an elliptical shape, the final shape as measured is a compromise between two ellipses both in direction and in axial ratio (Figure 8.3). The fluctuation in observed axial directions and axial ratios will thus have a systematic relationship to the tectonic axes.

Figure 8.4 shows a graphical method of finding the tectonic strain ratio and orientation of the major axis of the strain ellipse by plotting R, the observed strain ratio Y'/X', against α, the angle that the long axis of the ellipse X' makes with reference line α_0. If enough measurements are made, the plot yields maximum and minimum values of R (Figure 8.4A). Now since

$$R_{max}^2 = R_T R_0 \qquad (8.1)$$

and

$$R_{min}^2 = R_0/R_T \qquad (8.2)$$

where R_0 is the original ratio Y'_0/X'_0 and R_T the

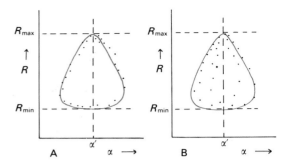

Figure 8.4 Finding the principal strain ratio in two dimensions using pebbles with an initial elliptical shape. A. Plot of $R = X'/Y'$ (observed ratio) against angle α with reference line for a constant initial shape ratio R_0. The principal tectonic strain ratio is found by taking the positions of R_{max} and R_{min} and using the relationship $R_T = R_{max}/R_{min}$ where R_T is the tectonic strain ratio (see text). The angle which X' makes with the reference line is α'. After Ramsay, J.G. (1967), *Folding and Fracturing of Rocks*, McGraw-Hill, New York, Figs. 5.27, 5.28.

tectonic strain ratio Y'_T/X'_T, the tectonic strain ratio may be found by dividing (8.1) by (8.2) giving

$$R_T = R_{max}/R_{min} \qquad (8.3)$$

If the original ratio R_0 is variable but smaller than R_T, the relationship still holds, but R_0 will represent the maximum value of the original ratio and the points on the graph will be distributed throughout the area bounded by the envelope shown in Figure 8.4B. If the orientation of the maximum strain axis X' is not already known, it is given by the symmetry axis of the plot (α' on Figure 8.4A, B).

8.4 Use of bilaterally symmetrical fossils

Another very useful method of strain determination employs the angular distortion of an original right angle in a bilaterally symmetrical fossil such as a brachiopod or a trilobite (Figure 8.5). The divergence from an original right angle will give the angular shear strain ψ. To obtain the ratio of the principal strains in the plane of the fossils, only one measurement of ψ is needed, provided that the orientations of the principal strain axes in the plane are already known. The axis of greatest extension X' is commonly visible as a '*stretching lineation*'. If θ is

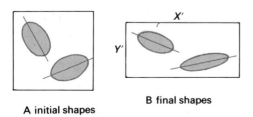

Figure 8.3 Effect of an initial elliptical shape on the strain ratio and orientation. A. Initial shapes. B. Final shapes after a homogeneous strain X', Y'.

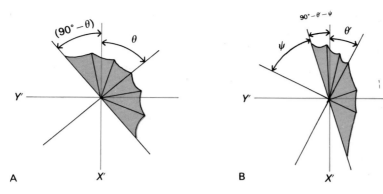

Figure 8.5 Finding the principal strain ratio in two dimensions using the angular shear strain in a bilaterally symmetrical fossil. A. Before strain. B. After strain. The angular shear strain is ψ and the strain ratio is given by $Y'/X' = \sqrt{(\tan\theta'/\tan(\theta' + \psi))}$—see text for explanation.

the original angle between the axis of symmetry of the fossil and X', then $90° - \theta$ is the complementary angle between the perpendicular to the symmetry axis and X'. Both these angles change as a result of the angular shear strain ψ: θ to θ' and $(90° - \theta)$ to $(90° - \theta' - \psi)$. From equation (6.5),

$$\frac{Y'}{X'} = \frac{\tan\theta'}{\tan\theta} \quad \text{and} \quad \frac{Y'}{X'} = \frac{\tan(90° - \theta' - \psi)}{\tan(90° - \theta)}$$

where X' and Y' are the principal strains in the plane concerned. Thus, multiplying,

$$\left(\frac{Y'}{X'}\right)^2 = \frac{\tan\theta'\tan(90° - \theta' - \psi)}{\tan\theta\tan(90° - \theta)}$$

$$= \tan\theta'\tan(90° - \theta' - \psi)$$

$$(\text{since } \tan\theta\tan(90° - \theta) = 1)$$

$$= \frac{\tan\theta'}{\tan(\theta' + \psi)} \quad (8.4)$$

'Short' and 'long' forms without angular shear strain. The above method is of course dependent on knowing the direction of X' and Y' in the plane of the fossil. Where these are not immediately obvious from the stretching lineation, they may be found by observing shortened or elongated forms of a deformed fossil that have suffered no angular shear strain (Figure 8.6). The axes of bilateral symmetry in these cases will be parallel to the maximum and minimum strain

axes in the plane and a simple comparison between them will often yield the strain ratio directly. Thus, assuming that the length/breadth ratio of both samples was originally identical, the strain ratio can be found as follows.

If l_1 and l_2 were the original lengths of forms A and B, and b_1 and b_2 their original breadths, respectively, then for A,

$$x_1 = l_1 X' \quad \text{and} \quad y_1 = b_1 Y'$$

and for b,

$$x_2 = b_2 X' \quad \text{and} \quad y_2 = l_2 Y'$$

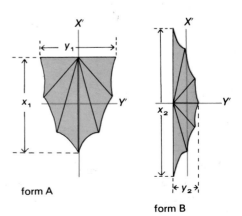

form A

form B

Figure 8.6 Finding the principal strain ratio in two dimensions using short and long fossil shapes showing no angular shear strain. If the lengths of the short and long forms are respectively x_1 and x_2 and the breadths y_1 and y_2, then the strain ratio $Y'/X' = \sqrt{(y_1 y_2/x_1 x_2)}$ (see text).

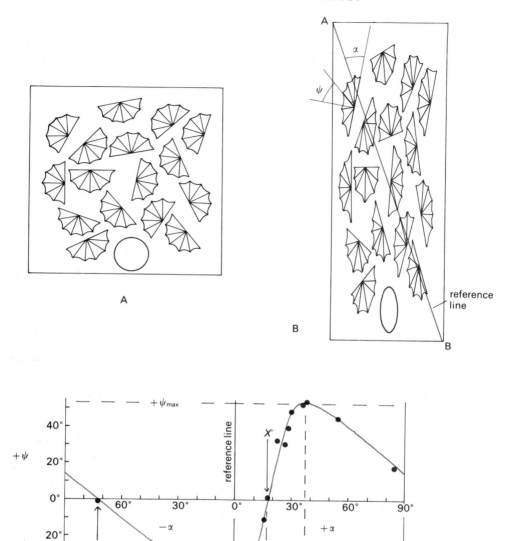

Figure 8.7 Finding the principal strain ratio in two dimensions using a number of bilaterally symmetrical fossils with varying orientation. *A, B*. The two diagrams show the initial and final orientations and shapes of the fossils after a homogeneous strain of Y'/X'. *C*. The values of α which the hinge lines of the fossil make with the reference line in *B* are plotted against the value of the angular shear strain ψ for each. The directions of Y' and X' with respect to the reference line are given at the points where the curve cuts the α axis, and the value of the strain ratio Y'/X' is θ' (see text). After Ramsay, J.G. (1967), *Folding and Fracturing of Rocks*, McGraw-Hill, New York, Figs. 5.60 and 5.62.

where x_1 and x_2 are the new lengths in the X' direction and y_1 and y_2 the new lengths in the Y' direction.

Thus, multiplying

$$\frac{y_1 y_2}{x_1 x_2} = \frac{b_1 Y' \cdot l_2 Y'}{l_1 X' \cdot b_2 X'}$$

$$\left(\frac{Y'}{X'}\right)^2 = \frac{l_1 b_2}{b_1 l_2} \cdot \frac{y_1 y_2}{x_1 x_2}$$

If we can assume that $b_1/l_1 = b_2/l_2$ and $(b_1 l_2)/(l_1 b_2) = 1$, then

$$\left(\frac{Y'}{X'}\right)^2 = \frac{y_1 y_2}{x_1 x_2} \qquad (8.5)$$

Several fossils showing angular shear strain. Where there are a number of fossils in the same plane showing variation in angular shear strain, it is possible to use a more accurate graphical method to find both the strain ratio and the orientation of the strain axes (Figure 8.7). It is not necessary that the fossils all be of the same species.

In this method the angular shear strain ψ (measured as in Figure 8.5) is plotted against the angle α between a standard identifiable line in the fossil (e.g. a hinge line) and an arbitrary reference line on the plane (Figure 8.7B). If the section shows sufficient variation in orientation, the graph will take the form of a curve cutting the α axis at two points which give the directions of the maximum and minimum strain axes (Figure 8.7C). The strain ratio is found by estimating the maximum angular shear strain from the curve and using the formula

$$Y'/X' = \tan\theta'/\tan\theta \qquad \text{(see equation 6.5)}$$

Since the maximum shear strain occurs when $\theta = 45°$ and $\tan\theta = 1$,

$$Y'/X' = \tan\theta'_{max} \qquad (8.6)$$

Alternatively, if the maximum or minimum points on the curve are insufficiently well defined, a number of estimates can be made using data for ψ and θ' for individual points using equation (8.4).

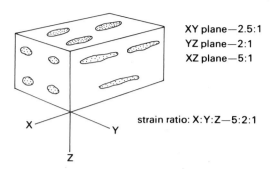

XY plane—2.5:1
YZ plane—2:1
XZ plane—5:1

strain ratio: X:Y:Z—5:2:1

Figure 8.8 Principal strain ratios in three dimensions. If the strain ratios are known for any two of the three principal strain planes (say XY and YZ) the strain ratios can be calculated in three dimensions.

8.5 Strain determination in three dimensions

The methods outlined in the preceding sections are designed to discover the strain ratio Y'/X' in a single plane. If this plane possesses two of the principal strain axes, e.g. X and Y, it is a comparatively simple matter to obtain a second strain ratio in a plane normal to the first, containing say Y and Z. The two ratios thus obtained, $Y:X$ and $Z:Y$, give a complete description of the three-dimensional strain. This is usually given in the form $X:Y:Z = 5:2.5:1$, for example, i.e. taking $Z = 1$ (Figure 8.8).

Where the positions of the principal strain axes X, Y and Z cannot be directly observed from the rock fabric, the strain may be determined using any three mutually perpendicular sections on which the strain ratio Y'/X' is measured. The method is described by Ramsay, J.G. (1967), *Folding and Fracturing of Rocks*, McGraw-Hill, New York, 142–147.

8.6 Use of fold sets

It is often convenient to measure strain as expressed in fold sets by comparing the final and (presumed) initial shapes of a layered sequence. Fold geometry in relation to the mechanism of formation is discussed in more detail in section 10. Briefly, the strain involved in folding may be analysed in two dimensions only, in the plane perpendicular to the fold axis, by comparing the initial and final lengths of a folded layer (Figure 8.9). If l_0 is the original length of a folded

Figure 8.9 Determination of shortening from a folded layer. The strain ratio $Y'/X' = l_1^2/l_0^2$ assuming the area of the layer hl_0 is constant.

layer of thickness h, and l_1 the new length, the *stretch* measured perpendicular to the axial plane is

$$Y' = l_1/l_0 \qquad (8.7)$$

The stretch parallel to the axial plane is found by assuming that the area hl_0 is constant and dividing by the new length, thus:

$$X' = \frac{hl_0}{l_1}\bigg/h = \frac{l_0}{l_1} \qquad (8.8)$$

The strain ratio is thus

$$\frac{Y'}{X'} = \frac{l_1^2}{l_0^2} \qquad (8.9)$$

This method takes no account of possible layer-shortening but this effect can be minimized by comparing several layers of different thickness and physical properties and taking the maximum value as an estimate of the total strain in that plane. A more serious restriction is that it can only measure two-dimensional strain and takes no account of strain parallel to the fold axes. On a large scale, however, estimates of strain based on the geometry of major folds are likely to be more accurate than observations based on strain markers from specific lithologies that may be biased in favour of ductile lithologies and more highly strained zones.

8.7 Two-dimensional strain from 'balanced sections'

Section 'balancing' has been developed as a method of unravelling complex thrust and exten-

sional fault zones by restoring them to their original lengths, in order to measure the fault displacement and to reconstruct the sequence of movements responsible for their often complex geometry.

The method is the same as that described in the previous section, except that, instead of considering a single layer, we are concerned with the change in dimensions of a planar section consisting of a number of layers. The area of the section is assumed to be constant, and the length l_1 and thickness t_1 of a deformed section are compared with the original 'stratigraphic' thickness t_0 of the same sequence in an undeformed section. The original length of the deformed section l_0 is then found by

$$l_0 = l_1 t_1/t_0 \qquad (8.10)$$

and the strain by

$$Y'/X' = l_1^2/l_0^2 = t_0^2/t_1^2 \quad \text{(from equation 8.9)}$$

The section must be taken perpendicular to the 'orogenic strike' (i.e. the main fold trend). Individual horizons in the deformed section are restored to their original lengths and the section is 'balanced' if the restored lengths of all the measured horizons are equal, and equivalent to l_0. Deformed sections are normally balanced down to a basal 'décollement' plane or sliding horizon below which the rocks are assumed to be unaffected by the shortening. Two common explanations for sections that do not balance are (i) inter-layer slip (very common in thrust tectonics) and (ii) different amounts of layer-parallel shortening (see Figure 10.3) in different beds (Figure 8.10). There is no unique solution to a section-balancing problem and usually some assumptions have to be made to achieve a solution. A common assumption is that restored thrust dips should be at 30°–35° with the horizontal. A three-dimensional study of fault displacements can be made if sufficient data are available and, except in the simplest types of fault belts, strain in the third dimension should be considered.

Figure 8.10 Construction of balanced sections. *A*. A section is restored by putting each layer back to its original position. The section 'balances' if all the restored layers are of equal length. Sections may not balance for several reasons. In *B*, the lower layers have been strongly distorted by the shear resistance at the ramp so that the displacement at the left side is 3 units compared with $5\frac{1}{2}$ units at the right side. Such distortion is often achieved by varying amounts of inter-layer slip. In *C*, the restored length of layers 1, 2 and 3 are different (*D*) because the lower layers have deformed more by shortening and thickening than by folding. Based on Dahlstrom, C.D.A. (1969) *Can. J. Earth Sci.* **6**, 743–747.

8.8 Bulk homogeneous strain

A quite different approach to the problem of strain measurement is to ignore the detail of local strain patterns, which are typically highly variable in any case, and treat the strain as statistically homogeneous over a given area. All the measured structural elements (foliations, fold axial planes, fold axes, etc.) in the area are plotted on a stereogram and compared with the ideal distributions that would be obtained with various types of strain ellipsoid. In this treatment we regard the bulk strain pattern as a large-scale strain ellipsoid, wherein planar and linear ele-

ments rotate passively in response to progressive strain towards the theoretical positions in the strain ellipsoid which would be taken up at infinite strain.

This model differs from the more familiar model of strain where we regard the planar structural elements as active, that is, the strain pattern is caused in part by the active rotation of planar surfaces around fold axes, for example. Instead we must imagine the bulk strain as if it were achieved by squeezing a collection of randomly-oriented planar markers situated within a ductile matrix.

The bulk homogeneous strain model is most

Figure 8.11 Stereographic plot (one quadrant) showing movement paths taken by poles to planes undergoing passive rotations during progressive homogeneous deformation (after Flinn, D. (1962) *Quart. J. geol. Soc. Lond.* **118**, 385–433). The dots show the positions of poles at infinite strain. *A*, $k = 0$, oblate strain; *B*, $k = 1$, plane strain; *C*, $k = \infty$, prolate strain (see text for explanation).

appropriate in studying areas of rather high strain where there was a considerable variation in attitude of pre-strain surfaces. These conditions apply particularly in certain Precambrian gneiss complexes that have been studied using this approach. Figures 8.11 and 8.12 show how the model may be applied.

Strain fields and movement paths. If we refer back to section 6.7, the various strain states were expressed in the Flinn diagram in terms of the value k. Figure 8.11 shows stereogram quadrants representing the three special cases, *oblate uniaxial strain* ($k = 0$), *plane strain at constant volume* ($k = 1$) and *prolate uniaxial strain* ($k = \infty$). The arrows show the movement paths taken by poles to planes which rotate passively under progressive strain. For oblate uniaxial strain, where $X = Y > Z$, planes will rotate towards the XY plane (the plane of flattening) and the poles will rotate towards Z; for plane strain, where $X > Y > Z$, again planes will rotate towards the XY plane but will first move towards XZ if their poles lie in or near XY and the poles again rotate towards Z; and for prolate uniaxial (constrictional) strain, where $X > Y = Z$, the poles to the planes will migrate towards the circumference of the stereogram and the planes will intersect along the X axis. As a consequence of the behaviour of planes under constrictional strains, fold axes will migrate into parallelism with X (since the fold axes represent the intersections of the planes of the fold surfaces). Directions of extension

indicated by boudinage structures (section 3.3) may also be plotted and will similarly migrate towards 'ideal' positions parallel to X, or into the XY plane where $X = Y$.

Measurement of homogeneous strain. In order to make use of these properties in providing a quantitative estimate of the bulk strain, it is necessary to measure the extent to which a strain ellipsoid departs from sphericity. The amount of deformation can be defined by r, where

$$r = a + b - 1 \qquad (8.9)$$

given that $a = X/Y = (1 + e_x)/(1 + e_y)$ and $b = Y/Z = (1 + e_y)/(1 + e_z)$ (see equations 6.12, 6.13).

Figure 8.12 shows the distribution of poles to planes in a stereogram at a strain level of $r = 3$ for a particular type of ellipsoid ($k = 1$). The stereogram is also contoured to show the limits of distribution of poles for different values of r.

Stereograms can be constructed for several different k-values. Each will have a different set of r contours, depending on the ratios of X/Y and Y/Z. The correct k-value could thus be found by comparison with these ideal distributions. Given a sufficient number of measurements and provided that our initial assumptions of random orientation are justified, it is possible to specify completely the shape of the strain ellipsoid and consequently to quantify the amount of strain.

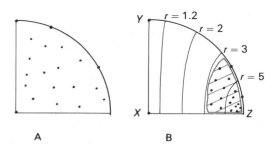

Figure 8.12 Equal-area plot (one quadrant) showing an initially random distribution of poles to planes (*A*), and (*B*) the distribution of the same poles at a strain level of $r = 3$ (where $r = a + b - 1$). Limits of distribution for various other r values are shown. Modified from Watterson, J. (1968) *Medd. om Grønland* **175**, Fig. 41.

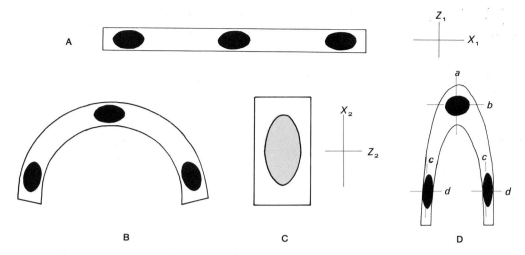

Figure 8.13 Addition of strains in two dimensions where the strain axes are parallel. *A* shows the initial strain X_1, Z_1 in a layer before folding. During folding (*B*) a second homogeneous strain X_2, Z_2 is added (*C*). The final strain (*D*) varies—at the hinge $a = Z_1 X_2$ and $b = X_1 Z_2$ whereas on the limbs $c = X_1 X_2$ and $d = Z_1 Z_2$ (see text).

8.9 Superimposition of strains

As for stresses, it is possible to add or subtract strains, and two strain ellipsoids of whatever orientation may be added to produce a 'compromise' ellipsoid. The principle may be illustrated by considering the effect of adding stretches θ_1, θ_2 and θ_3 of a strain ellipsoid *B* along the axes x, y and z of another strain ellipsoid, *A*, with principal strains X, Y and Z. The new stretches along x, y and z are obtained by multiplying the respective stretches, giving $X_1\theta_1$, $Y_2\theta_2$ and $Z_3\theta_3$. To determine the geometry of the new finite strain ellipsoid, the strains should be described in tensor notation as described by Means (1976).

Figure 8.13 illustrates the principle of superimposition of strain in two dimensions in the special case where the axes of the two strain ellipsoids correspond. The first strain is represented by three ellipses (with axes X_1Z_1) in the plane of the layer (Figure 8.13*A*). The layer is isoclinally folded (Figure 8.13*B*) and a second homogeneous flattening strain imposed, represented by an ellipse with axes X_2Z_2 (Figure 8.13*C*) giving the superimposed strain pattern of Figure 8.13*D*. The original strain ellipses are now oriented in such a way that $X_1 \parallel Z_2$ in the fold hinge whereas $X_1 \parallel X_2$ on the fold limb. The final strain in the hinge is therefore $a = Z_1 X_2, b = X_1 Z_2$ and in the limbs is $c = X_1 X_2$, $d = Z_1 Z_2$ giving final strain ratios in the hinge of $X_1 Z_2 / Z_1 X_2$ and in the limbs of $Z_1 Z_2 / X_1 X_2$.

Further reading

Hossack, J. (1979) The use of balanced cross-sections in the calculation of orogenic contraction: a review. *J. geol. Soc. Lond.* **136**, 705–711.

Means, W.D. (1976) *Stress and Strain*, Springer Verlag, New York.

Ramsay, J.G. and Huber, M.I. (1983) *The Techniques of Modern Structural Geology*, Vol. 1: *Strain Analysis*, Academic Press, New York. This work contains a comprehensive and rigorous treatment, with examples, of all the main methods of strain analysis. It is clear and easy to read, and is highly recommended to students who wish to pursue this topic.

9 Faulting

The morphology of faults has been described in section 1.1. In this section we shall discuss faulting as a process or mechanism, and how faults may be related to stress and strain.

9.1 Shear stress and brittle failure

Conditions for brittle failure. When a material fractures under deformation within its elastic

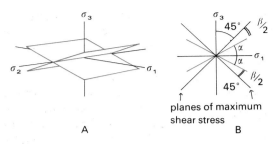

A

B

planes of maximum
shear stress

Figure 9.1 Relationship between shear fractures and principal stress axes. *A*. Shear fractures ideally intersect in σ_2 and make an acute angle with σ_1. *B*. Plane perpendicular to σ_2. Shear fractures make angle α with σ_1, and $\beta/2$ with the planes of maximum shear stress. Thus $2\alpha + \beta = 90°$.

range, it is said to exhibit *brittle failure* (section 7.3). The stress conditions at the point of failure are known as the *stress criteria of brittle strength*. These criteria include both the shear stress and the confining pressure, and vary with rock composition, temperature, etc.

When rocks fail under compression in experimental conditions, it is found that, in general, two sets of planar shear fractures are formed which intersect in a line parallel to the intermediate principal stress axis σ_2 (Figure 9.1*A*). Moreover the acute angle between the shear fractures is bisected by the maximum principal stress axis σ_1.

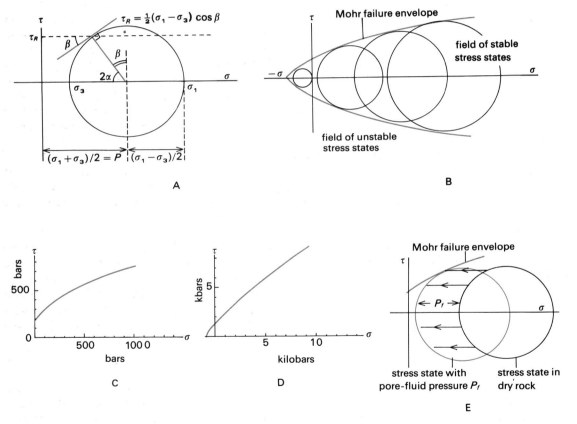

A

B

C

D

E

stress state with
pore-fluid pressure P_f

stress state in
dry rock

Figure 9.2 The Mohr stress diagram: failure criteria in two dimensions. *A*. Stress conditions at failure for a shear fracture making an angle α with σ_1 (see text for explanation). *B*. The Mohr failure envelope joining points of failure for different stress states separates the field of stable stress states from the failure field. Note that the shape of the Mohr curve implies an increase in the values of the shear stress τ and fracture angle α with increase in the mean (hydrostatic) stress $(\sigma_1 + \sigma_3)/2$. *C, D*. Mohr envelopes derived from the experimental deformation of Wombeyan marble (*C*) and Frederick diabase (dolerite) (*D*). After Paterson, M.G. (1958) *Bull. geol. Soc. Am.* **69**, 465–476, (*C*) and Brace, W.F. (1964) in *State of Stress in the Earth's Crust*, ed. W.R. Judd, Elsevier, Amsterdam, Fig. 22 (*D*), *E*. Effect of pore-fluid pressure. The black Mohr circle represents a stress state where the shear stress is too low for failure to occur. The effect of pore-fluid pressure P_f is to reduce the normal stress to the value represented by the coloured circle which intersects the failure envelope indicating that failure would occur.

The actual fracture planes do not correspond to the planes of maximum shear stress which make angles of 45° with σ_1 (see section 5.5). If the size of the acute angle between the fracture planes is 2α, then the difference between this angle and the 'ideal' angle made by the planes of maximum shear stress is $\beta = 90° - 2\alpha$ (Figure 9.1B). The angle β is sometimes referred to as the *angle of internal friction* of the material and varies for different stress states.

The Mohr stress diagram. This diagram (Figure 9.2A) is a convenient way of portraying the relationship in two dimensions between shear stress, confining pressure and angle of failure at the point when failure occurs. Each state of stress is represented by a circle with centre $(\sigma_1 + \sigma_3)/2$ and radius $(\sigma_1 - \sigma_3)/2$ that intersects the σ axis in two points σ_1 and σ_3. It is assumed for convenience that $\sigma_1 > \sigma_2 = \sigma_3$. If the values of stresses σ_1 and σ_3 at failure are represented by the circle and the angle between the shear fractures is 2α as shown in Figure 9.2A, the stress conditions at failure are represented by the point X. The shear stress at failure, τ_R, is given by

$$\tau_R = \tfrac{1}{2}(\sigma_1 - \sigma_3)\cos\beta \qquad (9.1)$$

from equation (5.4), since $\beta = (90° - 2\alpha)$. β is the angle that the tangent to the circle at X makes with the horizontal. The confining pressure at failure is given by

$$P = \tfrac{1}{2}(\sigma_1 + \sigma_3) \qquad (9.2)$$

and the normal stress across the fracture plane is given by

$$\sigma = \tfrac{1}{2}(\sigma_1 + \sigma_3) + \tfrac{1}{2}(\sigma_1 - \sigma_3)\sin\beta \qquad (9.3)$$

(from equation 5.3).

The effect of varying the stress conditions at failure is shown in Figure 9.2B, where a number of circles are drawn to show various values of σ_1 and σ_3 at which failure occurred for a particular rock specimen. The curve joining the points of failure for the different stress states is called the *Mohr failure envelope*, and divides the field of stable stress states within the envelope from the failure field outside. This curve illustrates the

general principle that the value of the shear stress τ required to produce failure increases as the confining pressure $(\sigma_1 + \sigma_3)/2$ increases; i.e. as the size of the Mohr circle increases.

The diagram also shows the effect of negative compressive stress, or *tensile stress*, represented by that part of the envelope to the left of the τ axis. It is clear that the value of shear stress required to produce failure under tension is much smaller than that under compression, which corresponds to the observation that rocks are much stronger under compression. Moreover, although most rocks have a finite tensile strength, their compressive strength is effectively infinite if the shear stress is below the required shear strength. It may also be observed from the shape of the Mohr envelope in the tensional field that tensile shear fractures make smaller angles with σ_1 (β is large), and that where the shear stress is zero, the value of σ gives the tensile strength of the material.

The rock composition has a marked effect on the general shape of the failure envelope. Figures 9.2C, D show two failure envelopes derived respectively from the experimental deformation of marble and diabase (dolerite). Note that the shear strength of the diabase increases more rapidly with increase in compressive stress than is the case for the marble.

Failure criteria. The shear stress τ acting along the fracture plane to promote failure is opposed by the compressive stress σ acting across the fracture plane which tends to close the crack and prevent failure (see Figure 5.3). The simplest relationship between shear stress σ and normal stress τ at failure is given by

$$\tau = c + \mu\sigma \qquad (9.4)$$

where c and μ (mu) are constants. This relationship is called the *Coulomb failure criterion* and gives a linear Mohr envelope with a slope of μ. However, very few materials behave in this way. A more realistic interpretation is given by the *Griffith failure criterion* which is based on the suggestion that failure results from the propagation and linking of minute defects ('Griffith cracks') in the material (cf. section 3.4). Stress concentration occurs around the ends of the cracks which spread spontaneously above a certain critical stress. The Griffith crack theory leads to the relationship

$$\tau^2 = |4\sigma_t(\sigma_t + \sigma)| \qquad (9.5)$$

where σ_t is the tensile strength (a negative value) of the material. This curve gives a parabolic Mohr envelope, which means that β is large at low confining pressures and small at high confining pressures, rather than constant as in the Coulomb criterion, and corresponds quite closely to many experimentally derived failure curves (cf. Figure 9.2C, D).

These criteria are based on the two-dimensional Mohr diagram assuming that $\sigma_2 = \sigma_3$. It is likely that, where $\sigma_1 > \sigma_2 > \sigma_3$, the value of σ_2 will have some effect on the criterion and therefore equation (9.5) can be modified to give a three-dimensional failure criterion of the form

$$\tau_{\text{oct}}^2 = |8\sigma_t P| \qquad (9.6)$$

where P is the confining pressure (hydrostatic stress) $\frac{1}{3}(\sigma_1 + \sigma_2 + \sigma_3)$ and τ_{oct} is the 'octahedral' or three-dimensional shear stress, such that:

$$\tau_{\text{oct}}^2 = \frac{1}{9}[(\sigma_1 - \sigma_2)^2 + (\sigma_2 - \sigma_3)^2 + (\sigma_3 - \sigma_1)^2] \quad (9.7)$$

This is referred to as the *Griffith–Murrell failure criterion*.

The Griffith–Murrell ('open crack') failure criterion is a close approximation to the behaviour of rocks failing under low confining pressures, but under high confining pressures where cracks would be closed by higher normal stresses, the Coulomb criterion is thought to be more accurate. For a fuller treatment of this subject, the reader is referred to Jaeger and Cook (1976).

Effect of pore-fluid pressure. As we saw in section 7.8, the effect of pore-fluid pressure is to reduce the effective confining pressure to a value $P_e = P - P_f$ where P_f is the pore-fluid pressure. For saturated rocks, where P_f approaches P, the value of the shear stress at failure becomes greatly reduced. The effect on failure can be shown by a simple change in the Coulomb criterion (equation 9.4) to

$$\tau = c + \mu(\sigma - P_f) \qquad (9.8)$$

which gives a reduction of μP_f in the value of the shear stress required to produce failure. The effect can be demonstrated on the Mohr diagram as a leftwards shift of the Mohr circle by an amount P_f (Figure 9.2E). If P_f is large enough, the circle will intersect the Mohr envelope and failure will occur, although the size of the shear stress may be much too low to produce failure in dry rock.

High pore-fluid pressures are particularly important in the movement of thrust sheets. To move large thrust sheets in dry conditions requires impossibly large forces to overcome the friction along the base of the sheet. If the thrust plane is lubricated by water, however, the required shear stress is reduced to a reasonable level. If we consider the problem of a thrust sheet sliding under gravity, the critical inclination of the plane of sliding should be about 30° in dry rocks, whereas much smaller inclinations are commonplace in nature. W.W. Rubey and M.K. Hubbert have studied the effect of pore-fluid pressure in reducing the critical slope necessary for gravity sliding. They show that, for example, the critical inclination may be reduced to only 5° if the pore-fluid pressure is 85% of the confining pressure. Pore-fluid pressures as high as this, and even higher, have been measured in deep boreholes.

9.2 Fault orientation in relation to stress and strain axes

We have seen that both theory and experimental results predict a simple relationship between the orientation of the principal stress axes and the shear fracture planes where σ_1 bisects the acute angle between two sets of shear fractures (Figure 9.1). This relationship can be used to investigate natural fault systems and in favourable cases to derive from them the orientation of the causal stress field. Since the theory applies to the initiation of fractures in completely homogeneous material, we would not expect very close agreement between theoretical and natural relationships, but nevertheless the theoretical model offers a very useful basis for understanding natural fault systems.

Since we can assume that shear stresses along

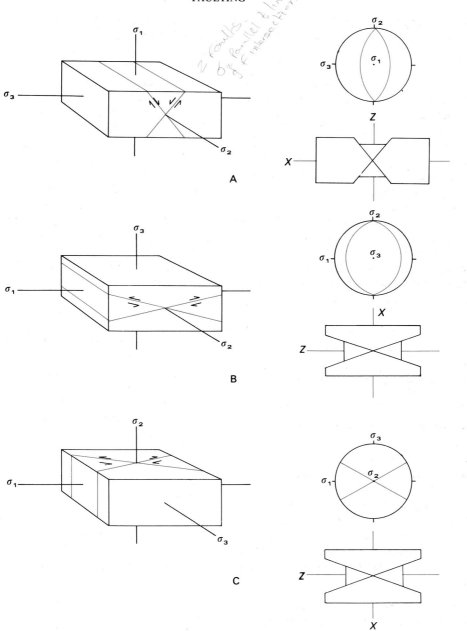

Figure 9.3 Fault orientation in relation to principal stress and strain axes. *A.* Normal fault sets. *B.* Thrust fault sets. *C.* Strike-slip fault sets. See text for further explanation. Stereograms, plan view. Strain diagrams, $\perp Y \parallel \sigma_2$.

the surface of the Earth are zero, it follows that one of the principal stress axes will be approximately vertical and the other two approximately horizontal. This leads to a simple threefold classification of fault sets based on the three possible orientations of the stress axes (Figure 9.3).

Normal fault sets (Figure 9.3A). Here σ_1 is vertical and corresponds to gravitational load. The two sets of normal faults intersect parallel to σ_2 and dip more steeply than 45°. Actual values of α (the angle between the fault plane and σ_1) are in the range 25°–30°. The sense of simple shear movement on these faults implies a contraction of the block parallel to σ_1 and extension parallel to σ_3 (Figure 9.4). Thus the X strain axis is parallel to σ_3, the Z axis is parallel to σ_1 and Y is unchanged (plane strain).

Thrust fault sets (Figure 9.3B). Here σ_1 is horizontal and σ_3 is vertical. Consequently two sets of thrust faults intersect along the horizontal σ_2 and dip less steeply than 45°. Actual values of α for thrusts are in the range 20°–25°. The sense of shear on the thrusts implies horizontal contraction parallel to σ_1 and vertical extension. It should be noted that symmetrical arrangement of thrusts as illustrated in Figure 9.3B is rare in nature; thrust sets are typically strongly asymmetric, with one direction dominant. Since the least stress must be vertical, thrusts will form more easily at relatively high levels in the crust where the lithostatic pressure is low.

Strike-slip fault sets (Figure 9.3C). Here σ_1 is again horizontal but, as σ_3 is also horizontal, two sets of strike-slip faults intersect along a vertical σ_2, each dipping vertically. Actual values of α for strike-slip faults are around 30°. The sense of shear on these faults implies horizontal contraction parallel to σ_1 and horizontal extension parallel to σ_3. As in thrust fault sets, one sense of movement is usually dominant.

Faulting in a non-homogeneous body. In a natural situation, where a stress field is applied to a rock body containing different rock types and planar discontinuities of different orientations, these simple rules break down. The effect of a plane of weakness (a previous fault, for example) oriented at the 'wrong' angle to σ_1 may be to cause faulting preferentially on that plane rather than to initiate a new fracture plane at the 'correct' angle. Both thrust and extensional fault systems (see sections 9.4 and 9.5) show examples of varying fault inclination (e.g. from thrust to reverse fault) controlled by movement rather than by initial stress state.

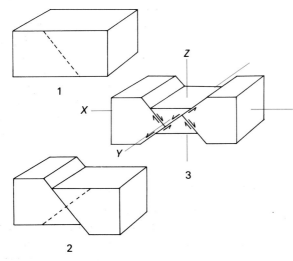

Figure 9.4 Shortening and extension of a block by complementary shear displacements on normal faults. This diagram explains the relationship between fault orientations and strain axes as shown in Fig. 9.3.

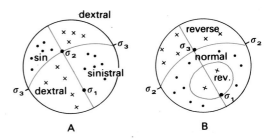

Figure 9.5 Finding approximate orientations of the principal stress axes from the sense of displacement on variably oriented faults. The principal stress planes shown on the stereogram divide faults with sinistral displacement from faults with dextral displacement (A) and faults with reverse displacement from faults with normal displacement (B). After Davidson, L. and Park, R.G. (1978) *Quart. J. geol. Soc. Lond.* **135**, 282–289, Fig. 6.

Where conjugate fault sets are found, the orientation of the principal stress axes may be determined by assuming that they intersect along σ_2 and that σ_1 bisects the acute angle, even if there is a certain amount of variation in the orientations of the faults. Where, however, there is no obvious conjugate set, and there is a range of orientations, any individual fault may make an angle with all three principal stress axes. In this case the direction of displacement on the fault plane depends both on the orientations and the relative sizes of the three principal stresses and cannot be simply interpreted (see section 5.5 and Figure 5.9).

Where the sense of displacement is known on a number of planes of various orientations, it is often possible to find the approximate positions of the principal stress axes. This is most easily done on a stereogram by plotting the positions of lines across which a change in sense of displacement occurs. These lines will contain the principal stress axes (Figure 9.5).

Stress trajectories and fault orientation. The simple relationship between the orientation of the stress axes and the horizontal, as shown in Figure 9.3, may hold true generally near the surface, but at depth, a variety of factors gives rise to variations in orientation of the stress axes which are reflected in turn in variations in the orientation of faults. In favourable circumstances it may be possible to use such a variation to draw *stress trajectories* (section 5.7) showing the variation in stress orientation which can then be related to theoretical models. Figure 9.6 shows

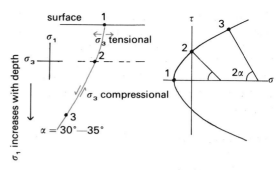

Figure 9.7 Change of fault inclination with depth of initiation due to increase in confining pressure, as predicted by the Mohr diagram (compare with Fig. 9.2). The hade of the fault (α) changes from $0°$ at point 1 (pure extension with no shear stress) to around $30°$ at point 3 within the compressional field. Note that faults initiated at a particular angle at depth may, in practice, propagate to the surface maintaining the same inclination; i.e. this diagram does not necessarily predict listric faults—see text for further explanation.

an example of a set of theoretical stress trajectories obtained for an elastic upwarp, together with a set of curved normal faults that would be predicted from such a stress distribution.

A variation in the angle at which a fault is initiated may be predicted, even under a uniform stress orientation, due to the effect of downward increase in confining pressure. This changes the fault angle α as shown by the change in the slope of the Mohr failure envelope (for example see Figure 9.2B) particularly where the horizontal stress changes from tensional to compressional with depth (Figure 9.7). Note that this effect need not result in curved (listric) faults. A fault initiated at depth at an angle of $30°$ may propagate to the surface without significant change in inclination. It is thought that most large faults probably initiate at depths of $10–15$ km in the strongest part of the crust.

9.3 Faulting and earthquakes

Fault initiation and propagation. When a brittle rock is compressed, certain strain effects take place before fracturing. These changes are important in earthquake prediction. The initial strain is elastic, but when the shear stress reaches a value of about half the shear strength, the rock begins to show some permanent strain due to the

Figure 9.6 Stress trajectories (coloured lines) and normal fault orientations (dashed lines) resulting from an elastic upwarp in two dimensions. Theoretical model, after Sanford, A.R. (1959) *Bull. geol. Soc. Am.* **76**, 19.

opening and propagation of small cracks in the zone of greatest strain in which the fault movement eventually occurs. The intensity of this microfracturing increases as the shear strength is approached. The opening of the cracks causes an increase in volume or *dilation* in the rock which is associated with an increase in the fluid content, as ground water migrates into the cracks. The rise in pore-fluid pressure has a significant effect in weakening the rock, as noted earlier (section 9.1).

The rate of propagation of the microfractures is the critical factor in determining the strain rate, and hence the time taken for fracturing (and the resulting earthquake) to occur. The influence of pore-fluid pressure on the strain rate has led to experiments at the Rangely Oil Field in Western Colorado, where the level of earthquake activity was artificially increased by pumping water into a fault zone. It has been proposed that large and potentially destructive earthquakes could be prevented by generating smaller ones by this method, and thus releasing the stress more safely.

Recognition of impending earthquakes. There are several features associated with the above changes which can be used to predict earthquakes. The dilation of the rock is accompanied by a decrease in *P*-wave seismic velocities and also by an uplift of the ground around the area of the fault. Another change that has been noticed is an increase in the amount of radon (an inert gas) in the atmosphere, presumably because of its release during the microfracturing process. The increase in pore-fluid pressure causes an increase in electrical resistivity which can be easily measured. All these changes are reversed in the period of rapid deformation leading up to the earthquake. Lastly, all large earthquakes appear to be preceded by a number of small shocks that increase in frequency immediately before the main earthquake.

Method of fault movement. When a fault plane has become established, further strain effects are in the form partly of very fast movements ('*slip*') along the fault plane, and partly of slow movements comparable to the pre-fracturing strains.

Figure 9.8 Lateral replacement of seismic slip by aseismic strain (see text for explanation).

The former are seismic, with displacement rates of the order of metres per second, whereas the latter are aseismic with velocities of the order of centimetres per year—similar to other aseismic crustal movements. Only a limited section of a large fault will take part in a particular episode of seismic slip, and the slip displacement dies out gradually at both ends into regions of aseismic movement (Figure 9.8). Any given section of the fault plane will therefore exhibit short periods of fast seismic slip separated by long periods when that part of the fracture is inactive—i.e. the fault 'sticks'. This behaviour is known as *stick-slip* and is typical of the region between 4 km and 10 km in depth, where the majority of fault-generated earthquake foci occur. Above 4 km, stable sliding takes place because the compressive stress across the fault plane is low. Below 10 km, because of the increase in confining pressure, there is a transition to a more ductile form of deformation (see section 9.5).

Secondary stress fields. The process of faulting, by locally releasing stress in the strained zone, and by the lateral movement of blocks of rock along the fault, causes a modification of the stress field around the active region which may in turn influence further fault movements. Secondary stress fields are particularly important around the end of a line of active slip. Figure 9.9 shows an example of how a new set of stress trajectories may be derived by superimposing new compressional and tensional stresses σ'_1, σ'_3 parallel to the fault on an oblique set of pre-faulting stress axes σ_1. Complicated systems of branching or *splay*

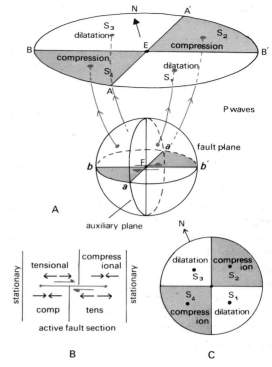

Figure 9.9 Stress trajectories around the end of an active fault. *A.* The initial set of σ_1 trajectories together with the new stresses σ_1' and σ_3' arising from the additional compressive and extensional effects due to the fault movement (see also Fig. 9.10*B*). *B.* The addition of the two sets of stresses produces a new set of stress trajectories—i.e. a combined stress field. After Chinnery, M.A. (1966) *Can. J. Earth Sci.* **3**, 163–174.

faults at the end of a major fault (see Figure 1.10) may be explained in this way.

Earthquake focal mechanism solutions. The orientation of fault planes and the displacement direction along them may be determined under favourable circumstances by a seismological method called *first-motion study*. This method is particularly useful in deciding on the origin of earthquakes relating to concealed faults, especially in the oceans, and has proved to be very important in the development of plate tectonic theory by enabling the relative motions of lithospheric plates to be determined.

This method is illustrated in Figure 9.10. If an earthquake originates by a shear displacement along a section of fault plane, the plane perpendicular to the displacement vector of the fault and midway along the displaced sector (the *auxiliary plane*) will divide regions of compression from regions of tension. These regions will be in

Figure 9.10 Focal mechanisms of an earthquake: first-motion study. *A.* A small sphere has been drawn around the focus of an earthquake *F* resulting from movement along an active segment of fault *b–b'*. The dextral fault movement results in compression and dilatation in opposite quadrants (see also *B*). The P waves originating in the compressive quadrants will show compressive first motions on arrival at the surface at recording stations S_2 and S_4, and those originating in the dilational quadrants will show dilational first motion on arrival at S_1 and S_3. Given enough stations, the orientation of the planes dividing the quadrants can be determined (*C*). For simplicity, in this example the fault movement is horizontal and the fault plane vertical, but oblique fault movements can be reconstructed using the same method. The effect of any oblique fault displacement can be shown by tilting the sphere to the required orientation. After Bolt, (1978) 67.

opposite quadrants since the movement is in opposite directions on each side of the fault (Figure 9.10*A*, *B*). The pattern of compression and dilation is preserved in the seismic waves that are radiated by the earthquake source and the phase of the initial seismic wave received at the recording station reflects its origin. Thus if there are a sufficient number of seismograph stations in different directions from the earthquake, the orientation of the planes dividing the compressional and tensional quadrants can be determined (Figure 9.10*C*). One of these planes is the fault plane and the other is the auxiliary plane. It is not possible from the first-motion study alone to determine which of these planes is the fault, but if one is more likely, given our knowledge of the local geology, then the direction and sense of movement can be obtained as shown, the displacement vector being perpendicular to the line of intersection of the planes.

9.4 Thrust systems

The detailed structure of many thrust belts is extremely complex (see for example the Moine thrust belt—section 15.6 and Figure 15.7). However, much of the complex interrelationship between faulting and folding may be explained by careful reconstruction following certain simple basic rules. The method of estimating shortening in thrust belts by balanced sections is described in section 8.7.

Thrust geometry. Where thrusts affect a set of bedded rocks that are near-horizontal in attitude, the thrusts generally follow a *staircase* path made up of alternating *ramps* and *flats* (Figure 9.11*A*). The flats are where the thrust sheet slides along a relatively weak bedding plane (often called a *décollement* plane) and the ramps are sections where the thrust cuts upwards through the stratigraphic sequence at an angle of typically around 30° to the horizontal. In uninverted strata, thrust displacements of this type always place older strata upon younger.

The ramp that is cut in the hangingwall is oblique to the bedding in the hangingwall sheet. Therefore, as this sheet moves up the footwall ramp and along a flat, the bedding will display an anticlinal structure (Figure 9.11*B*). Ramps do not necessarily strike perpendicular to the direction of transport, but are also found oblique or even parallel to the transport direction, in which case there will be a strike-slip component of movement along them.

Sequence of thrusting. Thrusts may develop in sequence either forwards or backwards from the first thrust (Figure 9.11*C, D*). Where the later thrusts develop in the footwall of the original thrusts, the earlier hangingwalls are carried forwards in a *piggyback* manner by the later (Figure 9.11*C*). Conversely, if the thrusts migrate backwards so that later thrusts develop in the hangingwalls of the earlier, an *overstep* sequence develops so that the higher thrusts will be the later (Figure 9.11*D*).

Formation of a duplex. Thrust sequences often result in the stacking up of many thrust sheets making up an *imbricate zone* or *schuppen structure*. In sequences of piggyback type, an imbricate zone may be bounded at the top by the original thrust surface, forming a *roof thrust*, and at the base by the currently active thrust surface or *sole thrust*, forming a *floor thrust* (Figure 9.11*E*). In this case the whole thrust package is termed a *duplex*. Individual imbricate sheets within the duplex are called *horses*. A typical duplex therefore consists of a roof thrust and a floor thrust enclosing a stacked-up pile of horses.

Back thrusting. Thrusts with a displacement direction, or *vergence*, opposite to that of the main thrust movement are known from many thrust belts. These *back thrusts* may be explained by the additional layer-parallel compression induced by gravity as the sheet climbs up the ramp. The uplifted hangingwall block thus formed is termed a *pop-up* (Figure 9.11*F*). If the back-thrust truncates an earlier thrust, a triangle zone is formed (Figure 9.11*G*).

Piggyback sequences are considered to be the normal mode of propagation in thrust systems, but many thrust belts exhibit local *out-of-sequence thrusts* that originate at or below the level of the sole thrust and cut up through the imbricate stack behind the thrust front, often using pre-existing fault planes. Such thrusts may locally violate the normal rule that thrusts place older strata upon younger. Another cause of complication in thrust belts is the occurrence of *low-angle normal faulting* due to extension at or near the thrust front. This has been attributed to

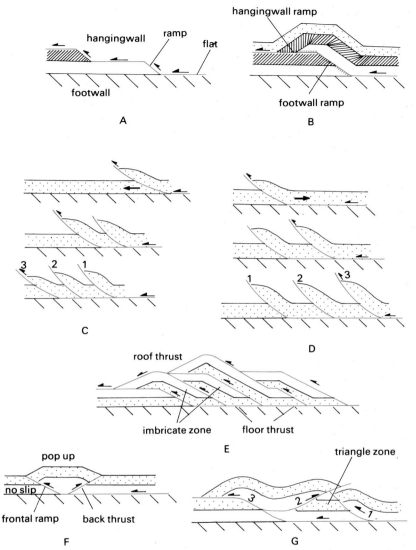

Figure 9.11 Structures and terminology of thrust zones. *A*. Shape of thrust surface: ramps and flats. *B*. Hangingwall geometry: a fold in the hangingwall must result from a ramp. *C*. Piggyback thrust sequence (new thrusts develop in the footwall). *D*. Overstep thrust sequence (new thrusts develop in the hangingwall). *E*. Structure of a duplex—imbricate thrust slices are contained between a floor thrust and a roof thrust. *F*. 'Pop-up' structure formed by backthrusting (see text). *G*. Triangle zone formed by backthrusting (see text). After Butler (1983).

gravitational sliding induced by the over-thickened thrust stack.

9.5 Extensional fault systems

It is comparatively common for regions of the crust to exhibit sets of related faults on which the individual displacements produce a net extension in the system as a whole. Such fault systems are characteristic of *extensional* or *divergent tectonic regimes*, which are associated (i) with constructive plate boundaries, at *oceanic* and *continental rift zones*; (ii) with destructive plate boundaries, in regions of *back-arc extension* such as the Basin-

Figure 9.12 Extensional fault systems. *A.* Method of accommodating extension by block rotation above a low-angle extensional fault-shear zone. After Wernicke, B. (1981) *Nature* **291**, 645–648. *B.* Listric fault with hangingwall roll-over anticline. Areas A and B are equal. The adjustment in hangingwall shape implies internal strain. *C.* Flat/ramp geometry of fault produces geometrically necessary folding in hangingwall. *D.* Antithetic faults in hangingwall. *E.* Synthetic faults in footwall. *F.* Extensional duplex with associated listric fan, counter fan and elevated central high. *B–F.* from Gibbs, A.D. (1984) *J. geol. Soc. Lond.* **141**, 609–620. *G.* Two *listric faults* linked by a *transfer fault*. All three faults detach on a sole thrust.

and-Range province of the western USA, and in oceanic marginal basins; (iii) with certain continental *intraplate* regions marked by rifts and extensional basins.

Classical views on faulting and on the formation of graben and rifts visualized extension being accommodated by dip-slip movements on steep normal faults (see Figure 9.3*A*) or by the filling of extensional fissures by magma. However, more recently, the importance of *low-angle normal faults* has been emphasized as a method of achieving the much larger extensions (of 100–200%) that have been estimated in certain extensional provinces such as, for example, the Basin-and-Range province.

There appear to be three principal geometric elements in *extensional fault systems*: (i) *low-angle normal faults* acting as detachment planes; (ii) *rotated fault blocks*; and (iii) curved or *listric fault* surfaces (Figure 9.12*A*). The widespread occurrence of such geometries has been confirmed by seismic records obtained in petroleum exploration of marine basins, and is summarized by Gibbs (1984). The listric fault achieves a rotation in the hangingwall as a geometric consequence of the displacement, and may be accompanied by an accommodation anticline, known as a *rollover anticline* (Figure 9.12*B*). The detachment or *sole fault* in extensional systems has the same function as that in thrust fault systems, and may possess a similar *ramp-flat* geometry (Figure 9.12*C*) necessitating complex accommodation strains in the hangingwall. The accommodation may take place by the formation of a set of *antithetic faults* that have the effect of extending and thinning the hangingwall in order to relieve the accommodation strain (Figure 9.12*D*). As extension proceeds, the sole fault may migrate into the footwall producing a set of *synthetic 'horsetail' listric faults*, known as a *listric fan* (Figure 9.12*E*), in an analogous way to the propagation of thrusts into the foreland in piggy-back thrust sequences. The fault blocks formed in this process are termed *riders*. The migration of the sole fault into the footwall may create an *extensional duplex* bounded by a *roof fault*. Figure 9.12*F* shows the complex type of geometry that may be produced by a combination

of hangingwall and footwall collapse resulting from progressive extension. The rotated blocks formed by these processes form *half-graben* at the surface; these may become filled with sediment. Upper-crustal extension on fault systems of this type may be transferred to mid-crustal levels and below by low-angle shear zones (see section 9.7).

Extensional fault systems also frequently contain steep faults with strike-slip displacements. Such faults are integral to the system and transfer displacement from one dip-slip plane to another. They are therefore termed *transfer faults*, and are smaller-scale counterparts of oceanic transform faults (see section 14.3). Figure 9.12*G* shows how two dip-slip faults may be linked by a transfer fault, each detaching on the same sole fault. Transfer faults may separate imbricate fault-fold packages that are geometrically distinct and uncorrelatable across the fault.

9.6 Strike-slip fault systems

Sets of related faults on which the individual displacements produce a net *strike-slip* displacement in the system as a whole are characteristic of tectonic regimes associated with major *conservative plate boundaries* (*transform fault* zones—see section 14.3) and also occur within continental plates such as, for example, the Asian plate north of the current plate boundary with the Indian plate (see Figure 15.5). The main features of a continental strike-slip regime have been analysed by Reading (1980) who pointed out the importance of differential vertical movements, which create uplifted blocks and depressed basins in such a zone.

Major strike-slip zones such as the San Andreas fault zone are of the order of 100 km across, within which quite complex tectonic effects take place. If the zone is considered as a deformable sheet, the deformation may be visualized as a combination of pure shear and simple shear, the pure shear component arising from net extension or compression across the zone, and the simple shear component from the strike-slip displacement of the boundaries of the zone. Considering only the simple shear component,

Figure 9.13 Strike-slip fault systems. *A.* Diagrammatic representation of the structural pattern produced by a dextral simple-shear couple, after Harding, T.P. and Lowell, J.D. (1974), *Bull. Am. Assoc. petrol. Geol.* **63**, 1016–1058 and Reading, (1980). *B.* Local compressional and extensional structures produced by fault terminations and fault overlaps in strike-slip faulting. *C.* Formation of raised and depressed wedge-shaped blocks by local transpression and transtension on a branching strike-slip system. *B, C.* after Reading (1980). *D.* Positive and negative *flower structures* produced by convergence and divergence respectively in strike-slip motion. Dot and cross symbols within circles indicate out-of-page and into-page components of motion respectively. *E.* Diagrams illustrating the formation of strike-slip duplex structure in transpression (A) and transtension (B). Note that the structures are analogous morphologically to compressional and extensional dip-slip fault duplexes respectively. *D, E.* from Park, R.G. (1988) *Geological Structures and Moving Plates*, Blackie, Glasgow and London.

the possible structures are summarized in Figure 9.13*A*, which shows the orientations of normal faults, reverse faults, and conjugate sets of strike-slip faults, as well as folds, within the zone. The orientation of these subsidiary structures within a strike-slip zone conveys important information as to the sense of strike-slip motion. The addition of an extensional component across the zone produces *transtension* and a compressional component produces *transpression*. Transtension results in clockwise rotation of the extension axis (in dextral shear) and transpression in anti-clockwise rotation (see Figure 9.13*B*). The reverse would hold for sinistral shear.

Important geometrical effects result from the irregularity of the fault geometry. Local changes of direction in strike-slip faults create local zones of transtension or transpression. In complex fault networks where branching faults have opposed dips, this process leads to alternate zones of raised and depressed blocks resulting from the convergence and divergence of wedge-shaped blocks (Figure 9.13*B*). The combinations of folds and faults produced in these local zones are termed *flower structures*. These are *positive* for uplifted blocks and *negative* for depressed blocks (Figure 9.13*D*). *Strike-slip duplexes* (Figure 9.13*E*) may be formed at bends in a strike-slip fault as a result of the progressive migration of the active fault into one wall.

The duplexes may be either transtensional or transpressional, and are morphologically equivalent to extensional and thrust duplexes respectively. Large pieces of crust may become detached and isolated from one block, transferred to the opposing block, and transported far from their sites of origin, becoming *displaced terranes* (see section 15.3).

Another cause of complication arises from *fault offsets*, which produce localized strains around the fault terminations that are either compressional or extensional, depending on the sense of movement (see Figure 15.3).

Major strike-slip zones, particularly those that constitute plate boundaries, are presumed to continue at depth as strike-slip shear zones (see section 9.7). Studies of deeply eroded Precambrian shield regions demonstrate the importance of major strike-slip shear zones at deep crustal levels. Other strike-slip zones may detach on sub-horizontal décollement zones within or at the base of the crust.

9.7 Shear zones

Brittle faulting is generally confined to the uppermost 10–15 km of the crust. Below this level, due to the change in physical properties of rocks brought about by the increase in temperature and confining pressure, brittle behaviour gradually gives way to ductile flow. If we examine the structure of high-grade metamorphic terrains, representing uplifted examples of deep crustal material, we find numerous examples of *shear zones* (see section 2.7), which represent zones of ductile displacement analogous to faults but without discrete fracture planes. The existence of such structures at depth has led to the realization that steep major faults or fault zones at the surface must be replaced by ductile shear zones at deeper levels in such a way that the total displacement of the blocks on either side is maintained (Figure 9.14).

Geometry of shear zones. Ideally, shear zones are contained between two parallel planar margins and are produced by simple shear stresses acting

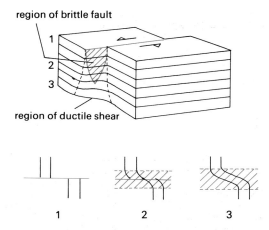

Figure 9.14 Transition from brittle fault to ductile shear at depth.

parallel to these margins (Figure 9.15). In practice, the margin may be very poorly defined since the strain near the margin is usually low (see Figure 9.17). An ideal shear zone represents *plane strain* and there is no displacement in the plane at right angles to the shear direction. The *displacement plane* therefore contains the X and Z principal strain axes and the Y strain axis remains unchanged. The *shear plane* contains the *shear direction*, and is parallel to the shear zone margins and perpendicular to the displacement plane.

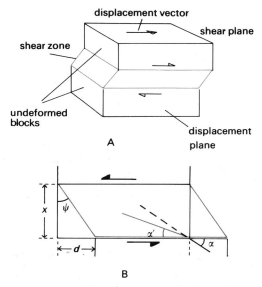

Figure 9.15 Main elements of shear zone geometry. *A*. The shear zone is a zone of ductile deformation between two undeformed blocks which have been displaced relative to each other. Note the shear direction (displacement vector) lying within the shear plane, which is parallel to the margins of the shear zone. *B*. The displacement *d* is given by $x \tan \psi = x \gamma$ where *x* is the width, ψ the angular shear strain and γ the shear strain. If a line makes an initial angle α with the shear plain and an angle α' after shearing, $y = \cot \alpha - \cot \alpha'$. After Ramsay (1980).

Shear zone displacement. There is a simple geometrical relationship betweeen *displacement*, *width* and *shear strain* in shear zones that is illustrated in Figure 9.15. If we take a section parallel to the displacement plane,

$$d = x \tan \psi = x\gamma \qquad (9.9)$$

where *d* is the displacement, *x* the width, ψ the deflection of a right angle and γ the shear strain (equation 6.4). In the general case of any line making an angle α with the shear plane, if the new angle is α'

$$\gamma = \tan (90° - \alpha') - \tan (90° - \alpha)$$
$$= \cot \alpha' - \cot \alpha \qquad (9.10)$$

Strain within shear zones. The mean shear strain within a shear zone can be simply calculated from equation (9.9), and equals the displacement *d* divided by the width *x*. This strain can be represented by a strain ellipse which makes an angle θ' with the shear plane (Figure 9.16*A*) such that

$$\tan 2\theta' = 2/\gamma \qquad (9.11)$$

(See Ramsay, 1980 for a detailed analysis of shear zone geometry.)

Normally the shear strain varies continuously across a shear zone, commencing with low values near the margins and reaching a maximum in the centre (Figure 9.16*B*). The variation in the angles θ' and α' with γ may be conveniently expressed graphically (Figure 9.16*C*). Where the displacement cannot be directly measured for a shear zone with varying shear strain, it can be calculated using the relationship

$$d = \int_0^{x_1} \gamma \, dx \qquad (9.12)$$

which can be calculated by plotting γ against *x* and measuring the area under the curve (Figure 9.16*D*).

Fabrics within shear zones. The variation of strain within shear zones is best studied by examining the fabrics developed in previously undeformed rocks, and many excellent examples may be found in deep-seated, coarse-grained, igneous rocks. Figure 9.17 shows a small shear zone in a gabbroic rock, where the strain is expressed by the deformation of originally equidimensional felsic and mafic grain aggregates into elliptical shapes, which become more

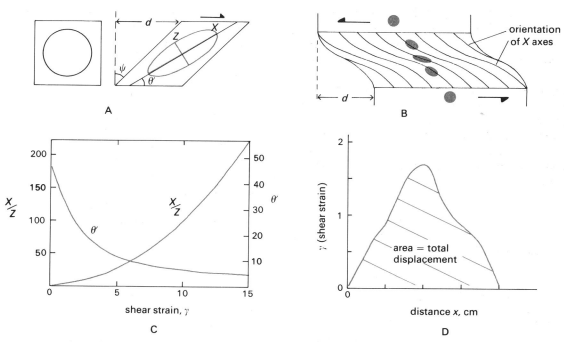

Figure 9.16 Shear strain within shear zones. *A.* Relationship between shear strain and strain axes (see text). *B.* Continuous variation in shear strain across a shear zone shown by a change in the orientation of the X strain axis and in the strain ratio X/Z. The angle θ' made by X with the shear direction is 45° at the margin of the shear zone and decreases to a minimum in the centre. *C.* Plot showing the variation of the angle θ' and the strain ratio X/Z with increase in the shear strain, γ. *D.* Plot of γ against x (distance across the shear zone) for the shear zone of *B.* The shear displacement d is equal to the area under the curve. *A–D* after Ramsay (1980).

Figure 9.17 Sigmoidal arrangement of fabric in a ductile shear zone in Lewisian metagabbro, N Uist, NW Scotland. Note that the fabric commences at an angle of 45° to the orientation of the shear zone and bends into near parallelism with the shear direction in the central part of the shear zone where it is most intensely developed. From Ramsay (1980), Fig. 2C.

elongate towards the centre of the shear zone. The long axes of these grain aggregates define a *shape fabric* or foliation that commences at the margins of the shear zone, making an angle of 45° with the shear direction. Towards the centre of the shear zone this fabric becomes more intense, and the angle that it makes with the shear direction decreases to a minimum. Clearly this fabric offers two methods of measuring the variation in shear strain: either plotting the variation of θ', or plotting the variation in the strain ratio Z/X of the deformed grain aggregates (see Figure 9.16C).

Termination of shear zones. Like faults, shear zones do not continue indefinitely. A shear zone may end at another shear zone, in which case the displacement is transferred to the other shear zone (see for example Figure 1.11 which shows how this can take place with faults). Alternatively, a shear zone may gradually die out into a wide zone of ductile deformation.

Further reading

Bolt, B.A. (1978) *Earthquakes: a Primer*, W.H. Freeman, San Francisco.

Butler, R.W.H. (1983) The terminology of structures in thrust belts. *J. struct. Geol.* **4**, 239–245.

Carreras, J., Cobbold, P.R., Ramsay, J.G. and White, S.H. (eds.) (1980) Shear zones in rocks. *J. struct. Geol.* **2**, 1–287.

Coward, M.P., Dewey, J.F. and Hancock, P.L. (eds.) (1987). *Continental Extensional Tectonics*. Spec. Publ. geol. Soc. Lond. **28**.

Gibbs, A.D. (1984) Structural evolution of extensional basin margins. *J. geol. Soc. Lond.* **141**, 609–620.

Jaeger, J.C. and Cook, N.G.W. (1976) *Fundamentals of Rock Mechanics*, Chapman and Hall, New York.

McClay, K.R. and Price, N.J. (eds.) (1981) *Thrust and Nappe Tectonics*. Spec. Publ. geol. Soc. Lond. **9**.

Price, N.J. (1966) *Fault and Joint Development in Brittle and Semi-brittle Rock*, Pergamon, Oxford. An excellent general textbook on faulting.

Ramsay, J.G. (1980). Shear zone geometry: a review. *J. struct. Geol.* **2**, 83–99.

Ramsay, J.G. and Huber, M.I. (1987) *The Techniques of Modern Structural Geology*, Vol. 2: *Folds and Fractures*, Academic Press, London.

Reading, H.G. (1980) Characteristics and recognition of strike-slip fault systems. *Spec. Publ. int. Assoc. Sedimentol.* **4**, 7–26.

10 Folding

10.1 Folding mechanisms and fold geometry

Several different mechanisms or methods of forming folds have been suggested. The most important distinction between these methods lies in whether a layer responds actively to a compressive stress applied parallel to its length in forming the fold, or whether it responds passively to a change in shape or in position brought about by movements taking place outside the layer and obliquely to it. The first of these methods may be illustrated by the *buckling* of a thin sheet under lateral pressure, and the second by the gravitational *bending* of a layer draped over a depressed basement block (Figure 10.1). Both are unrealistic models in the sense that they ignore the effect of the material enclosing the layer, which in practice has a very important role in determining or modifying the fold geometry.

Buckling. Ideally, in a fold produced by buckling under lateral compression of a single layer, the layer maintains its thickness throughout so that a *parallel* or *concentric fold* (see section 2.4) is produced. The strain produced within the layer is dictated by extension around the outer arc and compression in the inner arc, separated by a *neutral surface* of no strain near the centre of the layer (Figure 10.2A). The geometry of natural buckle folds is much more complex, however, and is discussed in section 10.3.

Flexural slip. This process involves a shear displacement or slip between successive layers deformed by buckling or some other mechanism (Figure 10.2B). This type of folding characterizes the deformation of relatively strong layers separated by planes or thin zones of weakness.

Figure 10.1 Active buckling (A) and passive bending (B).

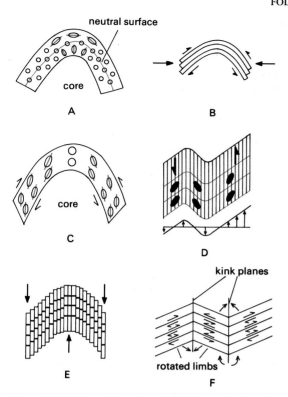

Figure 10.2 Fold mechanisms. *A.* Buckling showing the strain distribution within the folded layer. The neutral surface of no strain separates the extensional strain at the outer arc of the hinge from the compressional strain at the inner arc. *B.* Flexural slip—successive layers are displaced upwards towards the antiform crest with respect to the layer below. Individual layers are relatively unstrained. *C.* Flexural shear; the limbs of the buckle fold are modified by oppositely directed simple shear acting parallel to the layers. The hinge area is unstrained. *D.* Oblique shear (heterogeneous simple shear)—the fold is due to changes in the amount or direction of simple shear displacement. *E.* Card-deck model of an oblique-shear fold. *F.* Kinking—the fold is produced by the rotation of a set of layers on either side of a kink-plane (axial surface). The layers deform partly by flexural slip (see text).

Flexural shear. In this process, a fold produced by buckling, for example, is modified by simple shear acting parallel to the limbs of the fold, producing a strain distribution in which the long axes of the strain ellipses are divergent from the centre or core of the fold (Figure 10.2*C*).

Oblique shear or flow. If planes of simple shear are oblique or transverse to a layer and the amount and direction of shear displacement

varies along the length of the layer, a fold will be formed by passive rotation of the layer (Figure 10.2*D*). This process has been termed *heterogeneous simple shear.* The strain distribution is similar to that of flexural shear. This mechanism produces an ideal *similar fold* (section 2.4) and can be illustrated using a card-deck by drawing parallel lines on the edges of the deck and displacing the deck to make a fold shape (Figure 10.2*E*). The mechanism may also operate to modify the shape of an existing fold.

Kinking. This process forms folds of the *kink-band* or *chevron* type which typically have straight limbs and sharp hinges (section 2.4). The geometry is controlled by the rotation of sets of layers which remain planar between the kink planes, whereas rapid changes of orientation take place along the kink planes (Figure 10.2*F*). The limbs of the fold deform by flexural slip, and the process depends on ductile flow of material separating the stronger active layers. Kinking ideally produces folds of overall *similar* profile although individual layers exhibit different geometries (e.g. compare the white and black layers in the kink fold of Figure 10.21*B*).

Bending. The most common types of fold produced by bending are the accommodation structures formed in thrust and extensional systems, such as *hangingwall anticlines* and *rollovers* (see sections 9.4 and 9.5). In such folds the role of gravity is dominant, and the layer is deformed passively by accommodating to changes in shape of the surface over which it is moving. In practice, of course, the layers are also subject to layer-parallel compressive or extensional forces that result in the operation of other mechanisms such as buckling or kinking. Moreover the response of a layer to passive bending requires an additional flexural slip, flexural shear or flow mechanism, for example, to produce the change in shape.

Modifications due to homogeneous strain. The geometry of a fold is considerably modified if a homogeneous strain is imposed on a layer before, during or after the folding. Figure 10.3*A* shows the strain distribution produced by *layer shorten-*

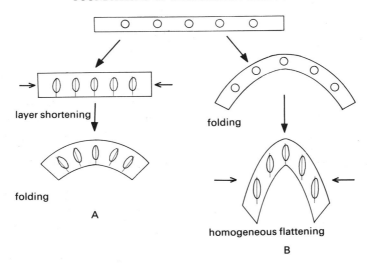

Figure 10.3 Strain pattern produced by (*A*) folding a layer shortened before folding, and (*B*) superimposing a homogeneous strain on a pre-existing fold.

ing prior to folding, and Figure 10.3*B* the strain distribution due to the superimposition of a *homogeneous flattening* strain on a fold. These strains have to be added to the strains produced by the folding process to give the total finite strain pattern.

Differences in geometry. Each of the ideal mechanisms described above produces a characteristically different fold geometry. These differences lead to a few simple geometrical tests which can be applied to natural folds in order to find out how closely they match one of the ideal types.

(i) *Ideal buckling* forms *parallel folds*; there is plane strain with the *Y* strain axis parallel to the fold axis and a combination of extensional and shortening strain in the hinge area. An initially straight lineation lying in the plane of the layer becomes curved during folding and the angle made with the fold axis remains unchanged only on the neutral surface (Figure 10.4*A*). On surfaces above and below the neutral surface, the lineation distribution changes, depending on the amount of strain.

(ii) *Ideal flexural shear* also produces *parallel folds* and plane strain, with *Y* parallel to the fold axis. The strain distribution defines a simple divergent fan of the *XY* planes. Since there is no distortion within the folded surface, the angle

made by a lineation within that surface, with the fold axis, is constant throughout the fold (Figure 10.4*A*) and the trace makes a small circle on a stereogram.

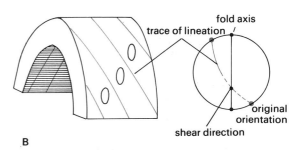

Figure 10.4 Simple models of lineation re-orientation during folding. *A.* Buckle folding of an initially straight lineation where there is no strain in the plane of the layer (e.g. at the neutral surface, or in flexural shear folding). *B.* Folding of an initially straight lineation during oblique-shear folding.

Figure 10.5 Oblique shear fold showing shear plane and shear direction. Note that the fold axis is oblique to Y.

(iii) *Ideal oblique heterogeneous shear* produces *similar folds* in which the thickness of the folded layer measured parallel to the shear plane is constant, but the orthogonal thickness varies systematically, thinning on the limbs. Again the process produces plane strain but the shear direction need not be perpendicular to the fold axis. Y is perpendicular to the shear direction and may or may not be parallel to the fold axis (Figure 10.5). An initially straight lineation is distorted in a systematic way such that it is rotated towards the shear direction. After folding, it therefore lies in a plane defined by the original orientation and the shear direction, that is, it forms a great circle distribution on the stereogram (Figure 10.4*B*).

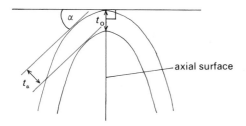

Figure 10.6 Method of measuring variation in layer thickness t_α in a fold. Measurement is made at the point where the tangent to the outer curve makes an angle α with the perpendicular to the axial surface. After Ramsay (1967), Fig. 7.18.

10.2 Classification of folds based on layer shape

A useful way of portraying the geometry of the fold profile, and thereby of comparing it with one of the ideal fold models, is to plot the change in thickness of the folded layer moving away from the hinge. The position of measurement is related to the angle α which the tangent to the curve makes at that point with the line perpendicular to the axial surface (Figure 10.6). The thickness of the layer at that angle, t_α, is then measured by taking the perpendicular distance between the two parallel tangents making the angle α to the inner and outer arcs respectively. A plot is made of the variation of $t' = t_\alpha/t_0$ with α (where t_0 is the thickness at the hinge). The curves of t' against α are characteristically different for the different fold models (Figure 10.7). The value of α reaches a maximum at the point of inflexion where the sense of curvature changes into the next fold. Measurements must be made in the XZ plane, which in the case of similar folds will not necessarily be perpendicular to the fold axis.

Dip isogon method. An alternative method of portraying the fold profile is to draw lines joining

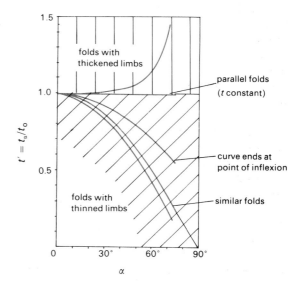

Figure 10.7 Graph showing curves of t' against α for various types of fold. $t' = t_\alpha/t_0$ where t_0 is the layer thickness at the hinge. After Ramsay (1967), Fig. 7.25.

points of equal limb dip (i.e. equal values of α) in successive layers through the fold profile. These lines of equal dip are called *dip isogons*. The method is illustrated in Figure 10.8. The dip of the isogons reflects differences in curvature between the outer and inner arcs of the folded layer. Thus if the isogons are *convergent* towards the core of the fold, the curvature of the outer arc is less than that of the inner arc. Conversely if the isogons are *divergent* towards the core of the fold, the curvature of the outer arc is greater than that of the inner arc. If the isogons are *parallel*, the curvature of the two arcs is equal (that is, the fold is *similar*). Isogons that are perpendicular to the folded layer indicate a *parallel fold*. These rules lead to a convenient fold classification.

Classification based on dip isogons. The fold classification is shown in Figure 10.9. There are three classes: *class 1*—folds with *convergent isogons*; *class 2*—folds with *parallel isogons* (*similar folds*); and *class 3*—folds with *divergent isogons*.

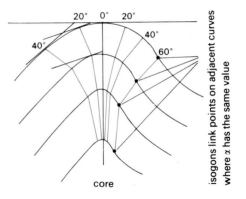

Figure 10.8 Construction of dip isogons. The isogons join points on successive fold surfaces with the same inclination α. In this example, the isogons are convergent towards the fold core. After Ramsay (1967), Fig. 7.18.

Class 1 is subdivided into three sub-classes according to the degree of convergence. Thus class 1A folds are strongly convergent, the isogons making an angle greater than α with the axial surface. Class 1B folds are *parallel folds*

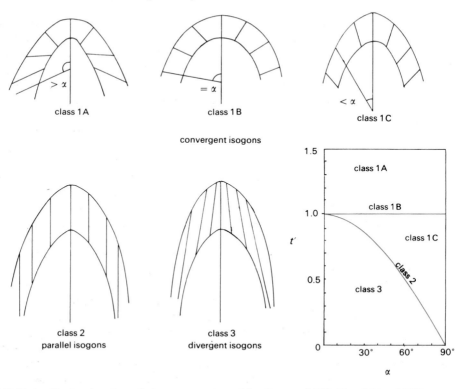

Figure 10.9 Fold classification based on dip isogons—see text. After Ramsay (1967), Figs. 7.24 and 7.25.

where the isogons are perpendicular to the fold surface and make an angle α with the axial surface. Class 1C folds are weakly convergent, the isogons making an angle less than α with the axial surface.

These classes can be shown on the t'/α plot (Figure 10.9). There is a large field of folds with thinned limbs, which is divided by the curve representing class 2 similar folds where the orthogonal thickness changes according to the relationship $t' = \cos \alpha$. It is rather important to measure such a fold precisely, since many similar-looking folds actually belong to class 1C and have probably formed by a different mechanism from the class 2 folds (see below).

10.3 Buckling

The term *buckling* is used to describe the process that produces a fold of approximately sinusoidal shape, or at least with relatively smooth changes of curvature, by compression acting parallel to the length of a layer. Folds formed in this way have a class 1 shape, often closely approximating to class 1B (parallel folds). Both experimental and theoretical studies have indicated that buckle folds will only form in relatively strong layers in a more ductile matrix and that there is a limiting value of the viscosity ratio between the layer and the matrix below which buckling cannot be initiated.

Controls on fold wavelength. Many natural sets of buckle folds exhibit a dominant wavelength that seems to be characteristic of a particular layer and is different from that of other layers in the same rock. Two key factors appear to control the wavelength: the *layer thickness* and the *viscosity ratio* between layer and matrix. The relationship between layer thickness and wavelength can be easily demonstrated by studying natural folds of layers of the same composition and varying thickness in a ductile matrix of constant composition (Figure 10.10A, B). The layers have to be sufficiently far apart for their waveforms not to interfere. Where they are close together, we find that a larger wavelength may be superimposed on a smaller as shown in

Figure 10.10C. Where buckling has affected layers of different viscosity as well, a more complicated arrangement results (see Figure 10.10D) which cannot be simply interpreted. Buckling of layers of different thickness or properties is one of the commonest causes of *disharmonic folding* (see section 2.5 and Figure 2.11).

Buckling of an interface. Buckle folding may also affect the planar interface between materials of contrasting viscosity. When this occurs, the folds have a characteristic form—those closing in one direction (e.g. the antiforms) have a broad rounded shape and those closing in the opposite direction (e.g. the synforms) have a narrow or *cuspate* shape (Figure 10.10E). The cusps always point towards the material with the higher viscosity. This is a common cause of *mullion structure* (see section 3.2 and Figure 3.6B).

Layer-parallel shortening in buckle folds. Comparing the lengths of folded layers of different wavelength gives us a method of estimating the amount of shortening that has taken place in the layers prior to buckling. Figure 10.12A shows an example of three layers where the present lengths (after unfolding) differ significantly. Layer 1, because of its greater thickness, has not folded. The minimum amount of layer-parallel shortening that has taken place is given by the difference in the lengths of layers 1 and 3. Thus by comparing the layers with the greatest and least unfolded lengths in a set of folds, an estimate of the layer-parallel shortening may be made. This is very useful in interpreting the strain distribution within the folded layers (see Figure 10.3A). Layer 2 will have experienced homogeneous strain before folding, whereas layer 3 may be almost unstrained before folding. Layer 1 might appear undeformed if the state of strain within the layer is ignored. It has been suggested that the amount of layer shortening is related to the viscosity contrast, in that the proportion of layer shortening to fold shortening increases with decrease in the viscosity ratio (see Figure 10.12B). With a low viscosity ratio, of course, no folding will occur and the layer will deform solely by layer shortening.

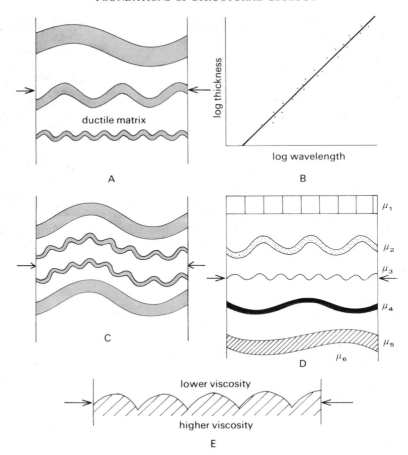

Figure 10.10 Influence of layer thickness and viscosity contrast on the wavelength of buckle folds. *A, B.* There is a linear relationship between log layer thickness and log buckle wavelength for widely separated layers of constant viscosity in a ductile matrix of much lower viscosity. *C.* Buckle folds of different wavelength may be superimposed if the layers are close enough to interfere. *D.* Disharmonic folding produced by a number of layers of different viscosity $\mu_1 - \mu_5$ and different thickness in a ductile matrix of much lower viscosity μ_6. After Ramberg, H. (1964) *Tectonophysics* **1**, 307–341. *E.* Buckle folding of an interface between two thick layers of contrasting viscosity. The cusps point towards the material of higher viscosity (cf. mullion structure, see section 3.2 and Fig. 3.6).

Because of the effect of shortening on layer thickness, the relationship between dominant wavelength and layer thickness described earlier requires modification. As deformation proceeds, there will be a tendency for folds with larger thickness/wavelength ratios to be favoured.

Shortening strain in parallel folds. The strain produced by parallel folding can be simply calculated by comparing the unfolded and folded lengths of a folded layer (section 8.6). The maxi-

mum shortening is achieved where the fold is concentric (Figure 10.13) in which case, given a layer thickness t, the initial length of the fold is πt and the new length $2t$. Thus the shortening e is given by:

$$e = (\pi t - 2t)/2t$$
$$= (\pi - 2)/2 = 0.36 (36\%) \qquad (10.1)$$

Further shortening strain can only take place by modifying the fold shape. This modification takes the form of a flattening of the fold perpendicular

Figure 10.11 *A.* Disharmonic folding, shown by different fold wavelengths in layers of different thicknesses, in finely laminated sandstone with interlayered mudstone. Note interference between closely-spaced layers with the same wavelength. New Harbour beds, Anglesey. *B.* Interference structure caused by superimposition of steeply inclined open to tight folds on earlier isoclinal folds in Precambrian banded gneiss, Härmanö, Orust, SW Sweden. *C.* Hand specimen of interbedded sandstone and slate showing marked cleavage refraction at the sandstone–slate boundaries and also bending of the cleavage in layers showing variable composition or grain size—e.g. the graded layers at the top of the photograph. Bendigo, Victoria. From Hills, E.S. (1972) *Elements of Structural Geology*, Chapman and Hall, New York, Fig. X-1.

to the axial surface and an equivalent extension parallel to the axial surface, and produces a fold with class 1C geometry known as a *flattened parallel fold*. The way in which a flattening strain is imposed on a buckle fold may be quite complex in detail, but in many cases the shape of the folded layer approximates closely to the result obtained if a homogeneous strain were simply added to the ideal parallel fold (Figure 10.14*A*). The addition of a homogeneous flattening strain with a strain ratio Z/X can be demonstrated on the t'/α plot of

Figure 10.7 as a family of curves of t'/α for various values of Z/X (Figure 10.14*B*)—that is, various degrees of flattening. The curves gradually approach the class 2 similar fold curve as the strain increases. When the limbs of an isoclinal fold become parallel and $\alpha = 90°$,

$$t' = Z/X \qquad (10.2)$$

That is, the strain ratio is given by the ratio between the thickness on the limbs and the thickness at the hinge. For a rapid estimation of

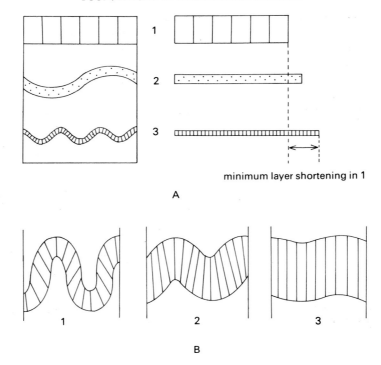

minimum layer shortening in 1

A

B

Figure 10.12 Layer-parallel shortening in folds of different wavelength. *A*. Unfolding of layers 1, 2 and 3 shows three different apparent original lengths due to different amounts of layer-parallel shortening before folding. The minimum shortening in layer 1 is shown (i.e. assuming that there has been no layer-parallel shortening in 3). *B*. Diagrams 1–3 show the effects of different viscosity contrast on the extent of layer-parallel shortening relative to fold shortening for 63.2% total shortening. Viscosity ratios—(1) 42.1, (2) 17.5, (3) 5.2. Experimental results from Dieterich, J.H. (1970) *Can. J. Earth Sci.* **7**, 467–476.

the amount of flattening in a non-isoclinal fold, it is sufficient to measure t_α/t_0 for two or three values of α and plot them on the graph of Figure 10.14*B*.

Strain within and outside the folded layer. The strain distribution within a buckled layer de-

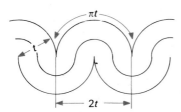

Figure 10.13 Maximum shortening in a concentric fold. The shortening is given by $e = (\pi t - 2t)/2t = (\pi - 2)/2$ where πt is the original length and $2t$ the new length of a folded layer. After Ramsay (1967), Fig. 7.48

pends on the actual mechanism involved (see Figure 10.2). Buckling that produces a neutral surface (Figure 10.2*A*) shows a type of strain distribution termed *tangential longitudinal strain*, where the strain axes are parallel to the layer boundaries. For certain types of fold profile, there is a concentration of strain in the hinge zones, but the limbs may be relatively unstrained. Where there has been some flexural shear (Figure 10.2*C*) the strain ellipses (and any associated planar fabric) have a *divergent fan* arrangement. Many natural sets of folds involve alternating layers of contrasting viscosity, such as sandstone bands in shale, for example (Figure 10.15). Where these deform by compression parallel to the layering, the more competent sandstone layers buckle to give, at least initially, a *convergent fan* arrangement of strain axes due to the predominance of tangential

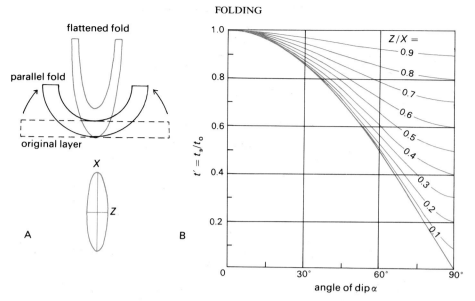

Figure 10.14 Superimposition of homogeneous flattening on a parallel fold. *A.* Homogeneous strain X, Z is superimposed on the parallel fold to produce a flattened parallel fold. *B.* Curves of t' against α for the determination of amount of flattening Z'/X' in buckle folds. After Ramsay (1967), Fig. 7.79.

strain. The more ductile shale, however, deforms rather differently. Where a narrow band of shale is constrained by the movement of adjacent competent layers, it will exhibit flexural shear which will produce a divergent fan of strain axes. However, away from the influence of active buckling layers, it will tend to deform more by homogeneous flattening, producing parallel strain axes.

These differences in strain pattern are very easy to detect where the rock contains a penetrative planar fabric, e.g. a slaty cleavage, that is parallel to the XY plane. The layers a–e in Figure 10.15 illustrate the variation in attitude of a slaty cleavage under these conditions. Such alternation of convergent and divergent cleavage fans is an example of *cleavage refraction*, which is a change in the attitude of the cleavage on passing from one bed into another of different composition. Within a competent layer, the cleavage planes may be curved. This is usually associated with a change in physical properties and is particularly common in graded beds where the degree of convergence decreases upwards with decrease in grain size (Figure 10.11C).

The pattern of alternating convergent and divergent cleavage fans is duplicated by the isogon geometry. Layers a, c and e of Figure 10.15 exhibit class 1C geometry, and layer b and the outer parts of layer d, class 3 geometry.

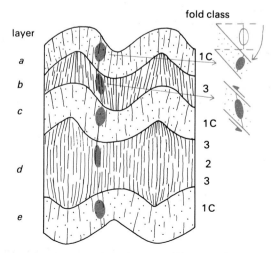

Figure 10.15 Cleavage refraction. Alternating convergent and divergent slaty cleavage fans produced by the folding of alternate competent and incompetent layers (see text).

The central part of layer *d* exhibits class 2 geometry. It is interesting to note that alternations of this kind may result in an overall class 2 geometry—that is, the dip isogons drawn from the top of layer *a* to the base of layer *e* would be approximately parallel, enabling the folding to continue indefinitely downwards.

In this way the individual buckled layers, with their rather complex strain geometry, when viewed on a larger scale, may become part of a much more homogeneous-looking system where the total shortening strains measured at various levels through the folded rock are approximately the same.

10.4 Oblique shear or flow folding

Folds may be formed by the passive rotation of a layer under simple shear directed obliquely to the layer. This process is often called *flow folding*, because of the analogy with the folding of marker layers in a flowing liquid. If the shear direction is constant through the fold, an ideal *similar* shape (class 2) is produced. In order to form a fold, the amount of shear displacement must vary along the layer, otherwise the layer would remain straight. Consequently the process is often called *heterogeneous simple shear*. The layers must behave passively for ideal class 2 geometry; if there is any component of buckling (that is, if the layers bend under lateral stress), the geometry will correspond to class 1 C, closely approaching class 2, perhaps, but never reaching it.

Folds of similar type only form in rocks which are in a very ductile state during deformation, typically under medium to high-grade metamorphism, although rocks of very low viscosity (such as salt and gypsum) behave in this way even at low temperature. Another essential condition is that the viscosity contrast between adjacent layers should be low—below the limit for the initiation of buckling.

Role of homogeneous strain. The imposition of a uniform homogeneous flattening strain on similar folds formed by heterogeneous simple shear does not change the similar geometry, provided that one of the principal strain planes of the homogeneous deformation is parallel to the shear plane. Given the ductile nature of rocks exhibiting such folds, it is likely that both mechanisms have operated together in most cases. The respective contributions of the two types of strain are illustrated in Figure 10.16. The simple shear strain only affects the fold limbs and produces a divergent fan of *XZ* planes. The addition of a suitably oriented homogeneous component increases the strain on the limbs, decreases the divergence, and produces a strain in the hinge zone equivalent to the homogeneous strain ellipse.

Origin of heterogeneous simple shear. The most likely cause of the heterogeneity in the simple shear component in most similar folds is the presence of a fold shape generated by a buckling

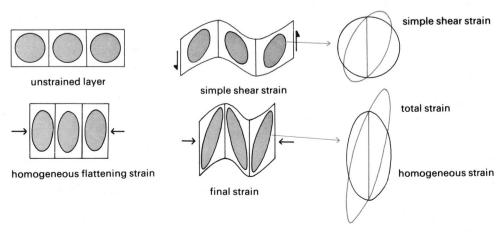

Figure 10.16 Superimposition of heterogeneous simple shear strain and homogeneous flattening strain in a similar fold.

or bending mechanism. If we imagine a thick ductile layer bounded by two stronger, more competent layers, the buckling of the competent layers will transmit their shape through the ductile layer as a heterogeneous simple shear. Although the strain in the ductile layer near its margins may be influenced by the buckling mechanism, a much more uniform pattern would be expected in the central part of the layer (see Figure 10.15*d*). The fact that an overall similar fold shape is retained in a pile of folds exhibiting alternating class 1C and class 3 shapes suggests that this process is important in generating similar folds.

Superimposition of oblique shear (flow) folds. It is geometrically much easier to deform an already folded surface by oblique shear than by buckling, since the shape of the passive layers is irrelevant and exercises no control over the second fold shape. The principle of superimposition of flow folds can be demonstrated by drawing a first-fold shape on a card deck and then deforming the card deck by heterogeneous shear (Figure 10.17).

Complex interference patterns are generated by superimposing a set of flow folds on a previously folded surface. The geometrical pattern depends on the spatial relationships between the two fold sets; in particular between the attitudes of the two fold axes and axial planes, and on the shear or flow direction of the second folds. Figure 2.14*B* (section 2.6) shows three common types of interference pattern produced on a flat surface by superimposition of fold sets—

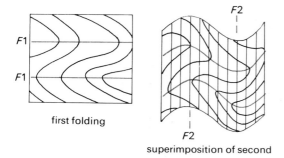

first folding

superimposition of second

Figure 10.17 Superimposition of oblique shear folds (flow folds) on a previously folded surface demonstrated by a card-deck model.

dome and basin, crescent and mushroom, and *double zig-zag.* These three types can be regarded as 'end members' in a continuous series of interference shapes. We may add a fourth type which produces no interference because the axial planes and axes of the two fold sets are parallel.

We can demonstrate how interference structures are formed by examining three cases (Figure 10.18). In each of these cases the second folds are upright with a vertical flow or shear direction. In the first case (Figure 10.18*A*) both the axial planes and axes of the *F*2 folds are near-perpendicular to those of *F*1, and the flow direction of *F*2 lies within or near the axial plane of *F*1, which is also upright. This type of relationship produces a dome and basin pattern.

In the second (Figure 10.18*B*) the *F*1 folds are recumbent and the axial planes and axes of *F*2 again make a large angle with those of *F*1. However, the flow direction of *F*2 also makes a large angle with the *F*1 axial planes. This relationship produces a crescent-shaped interference pattern.

In the third (Figure 10.18*C*) the *F*1 axial planes are gently inclined and thus again make a large angle with the *F*2 axial planes. However, the fold axes of both sets of folds are sub-parallel. The *F*2 flow direction is oblique to the *F*1 axial planes. This relationship produces a double zig-zag pattern.

As an example of how a mapped interference structure may be interpreted, let us return to the Loch Monar structure illustrated in Figure 2.14*C*. A three-dimensional model of this structure (Figure 10.19) shows that the plunge of the second folds is controlled by the attitude of the first fold limbs as well as by the *F*2 axial planes. Thus the *F*2 folds plunge vertically on the southern limb of the *F*1 synform, sub-horizontally across the hinge zone, and have a moderate plunge to the southwest on the northern limb. If we imagine the *F*2 folds to be unfolded, the shape of the *F*1 fold is revealed as an east–west-trending synform with a moderately south-dipping northern limb, a vertical southern limb and a low plunge to the west.

Formation of sheath folds. At high levels of strain,

1 *F*1 folds

2 *F*2 folds

3 interference

                               A   B   C

Figure 10.18 Interference patterns generated by the superimposition of upright flow folds on previous folds of varying attitude (see text for details). After Ramsay (1967), Figs. 10.3, 10.8 and 10.15.

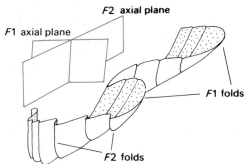

Figure 10.19 Geometry of an interference structure in the Moine complex at Loch Monar, Scotland. After Ramsay (1967), Fig. 10.25 (see text for details).

in shear zones for example, very complex fold shapes may be produced by a heterogeneous simple shear or flow superimposed on very simple buckles. During progressive simple shear, the fold axes may behave passively and rotate towards the shear direction until at high strain ($\gamma \simeq 15$) they will become sub-parallel to the shear direction (Figure 10.20). In this way it is possible to produce folds with strongly curved axes that ultimately develop very elongate dome and basin shapes. Such folds, termed *sheath folds*, are characteristic of strongly deformed parts of shear zones.

Figure 10.20 Development of a sheath fold by the progressive simple shear of an initial non-cylindrical fold. After Ramsay, J.G. (1980) *J. Struct. geol.* **2**, 83–99, Fig. 16.

10.5 Kinking and formation of chevron folds

The characteristic features of *kink bands* and *chevron folds* (see section 2.4) are straight limbs and sharp angular hinges. Kink bands (Figure 10.21A) are asymmetric structures where the deformation is essentially confined to the material within the kink band, whereas chevron folds are more symmetrical and continuous (Figure 10.21B, C). The mechanism, which appears to be responsible for both kink bands and chevron folds, is a process of flexural slip between the competent layers, which maintain their thickness, and ductile flow in the intervening incompetent material. This leads to folds with alternating class 1C and class 3 geometry which enables the folds to maintain an overall similar form.

The basic geometry of a kink band, as shown in Figure 10.21A, can be explained in terms of a rotation of the short limb of the asymmetrical fold couple so that the kink band behaves as a zone of simple shear. The rotation can take place with very little internal strain in the layers, provided that inter-layer slip can take place as shown. The amount of rotation of the short limb is expressed by the angle β. The kink plane makes an angle β_1 with the short limb and β_2 with the long limb. If $\beta_1 > \beta_2$ and the layer thickness remains constant, dilation must take place between the layers within the kink band. In nature this results in extensional fissures which are often filled with quartz or calcite (Figure 10.22A). Usually however, the kink layers rotate until $\beta_1 = \beta_2$ when no fissure space exists, and further

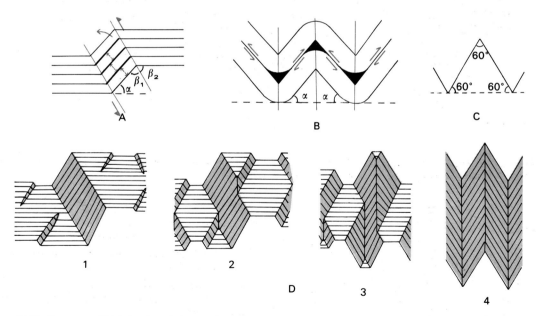

Figure 10.21 Geometry of kink bands and chevron folds. *A.* Kink band—the inclination of the short (rotated) limb is α and the kink plane makes an angle β_1 with the short limb and β_2 with the long limb. If $\beta_1 > \beta_2$, dilation occurs between the layers of the short limb. *B.* Chevron fold—this is a symmetrical structure where the limb inclinations α are equal. If the layers maintain a constant thickness, gaps occur at the hinges. *C.* An ideal chevron fold possesses limb inclinations of 60° and therefore also an inter-limb angle of 60°. *D.* Progressive growth of conjugate kink bands ultimately forming symmetric chevron folds. (1) 12.5%, (2) 25%, (3) 40%, (4) 50% shortening. From experiments on deformation of phyllite by M.S. Paterson and L.E. Weiss, *A, B, D* after Ramsay (1967), Figs. 7.111, 7.120 and 7.124.

Figure 10.22 Kink bands and chevron folds. *A*. Dilational veins (white) developed in a kink band in calcareous slates, Lake District. From Ramsay (1967), Fig. 7.119. *B*. Conjugate kink bands intersecting in a symmetrical box fold. West Cork, SW Ireland. From Ramsay (1967), Fig. 7.107. *C*. Chevron folds in Devonian phyllites, Boscastle, Cornwall. Note that ideal chevron folds should have straight limbs, whereas these are curved in places. It is common to find gradations between ideal chevron geometry and more rounded profiles. *A, B*, from Ramsay (1967). By permission, McGraw-Hill, New York.

movement would result in a contraction across the layers.

Kink bands usually occur in *conjugate* pairs or sets (Figure 10.22B) but to achieve maximum shortening by kinking, the whole length of the layers must be involved in the folding. Experiments on the compression of phyllite have shown that, under continued compression, individual kinks grow by lateral expansion until neighbouring oppositely directed pairs merge completely to form *chevron folds* (Figure 10.22C). This process is illustrated in Figure 10.21D. The development of chevron folds from kink bands will only take place when $\beta_1 = \beta_2$, that is, when the axial surface of the kink band (or kink plane) bisects the fold angle, and if $\alpha = \beta_1 = \beta_2 = 60°$. 'Ideal' chevron folds which are produced from kink bands with this geometry will thus have a fold angle of 60° (Figure 10.21C). Not all chevron folds may form in this way, however.

Shortening and internal strain in kink folding. Ideally the kinking process involves no internal strain in the layers on the limbs of the kinks and the strain is concentrated in the zone where the layer is bent around the hinge (Figure 10.23). The proportion of strained to unstrained layer is clearly greater when the layers are thicker in relation to the length of the limbs (compare Figure 10.23A and B). Thus the ratio between layer thickness t and limb length l is critical in determining the proportion of strained to unstrained material in the folds—e.g. the ratio of internal strain to total shortening is much greater in A than in B.

The amount of shortening that can be achieved by chevron folding depends on the angle α, but cannot be increased beyond a limiting value because of the 'locking' effect brought about by the progressive increase in internal strain. Actual 'locking values' of α vary with the t/l ratio and the

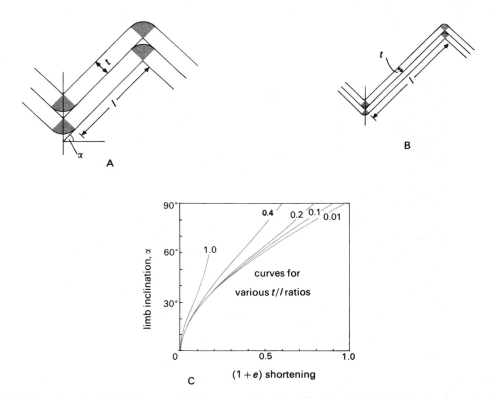

Figure 10.23 Internal strain in kinked layers of thickness t as a function of limb length l and limb inclination α. A. A high t/l ratio produces a large strained zone. B. A low t/l ratio produces a proportionally much smaller strained zone. C. Plot of shortening $(1 + e)$ against limb inclination for various values of t/l (see text). After Ramsay (1967) Fig. 7.112.

nature of the material. The relationship between t/l ratio, amount of shortening and α are shown in Figure 10.23C. This diagram shows that a fold with a low t/l ratio can accommodate a given strain with a smaller limb dip (or fold angle) than one with a high t/l ratio. We should expect therefore that chevron folds would form preferentially in thin-layered rocks. We find that in practice this t/l ratio is generally less than 0.1, associated with shortening strains of 50%–65%. Thus chevron folding is a more efficient method of shortening layers than buckling, since ideal buckles involve a maximum shortening of only 36% (see Figure 10.13).

Further shortening, as in buckle folds, will be accommodated by a modification of the chevron geometry by flattening of the limbs and extension of the hinges and can be approximated by the addition of a homogeneous strain (cf. Figure 10.14).

10.6 Conditions controlling the fold mechanism

There is no satisfactory theory at present that enables us to predict which of the three mechanisms—buckling, flow or kinking—will affect a given layered rock material under compression. However, there are certain generalizations which we may use as a guide. Firstly, it is clear that a *low viscosity contrast* between adjacent layers inhibits both buckling and kinking but favours flow. Secondly, a *high viscosity contrast* between adjacent layers is necessary for both buckling and kinking, but buckling of an individual layer or set of layers will be inhibited if the nearest competent layers are too close. Thus we would expect kinking to take place in a relatively thick sequence of thin closely-spaced competent layers, whereas we would expect buckling where the competent layers are separated by thicker layers of more ductile material. *Flexural slip* and *flexural shear* are mechanisms that may operate in either buckling or kinking processes.

The commonest type of occurrence attributable to *passive bending* is the accommodation folds associated with thrust and extensional fault systems (see Figures 9.11, 9.12). Such folds are produced essentially by the action of gravity as a layer accommodates to a change in shape caused by sliding of a sheet over a surface of changing inclination. The layering will then typically be subject to compression or extension in addition. The bending process clearly requires some additional mechanism or mechanisms to enable the layering to change shape. These may include flexural slip, flexural shear or oblique shear; if the layers are compressed, buckling or kinking may also be involved.

Finally we must remember that the scale of observation is very important in deciding both geometry and mechanism. Thus a layer which might be regarded as a buckle fold or a flow fold on a large scale might well involve elements of different mechanisms on a smaller scale.

Further reading

Cobbold, P.R. and Quinquis, H. (1980) Development of sheath folds in shear regimes. *J. struct. Geol.* **2**, 119–126.

Ramsay, J.G. (1967) *Folding and Fracturing of Rocks*, McGraw-Hill, New York.

Ramsay, J.G. and Huber, M.I. (1987) *The Techniques of Modern Structural Geology*, Vol. 2: *Folds and Fractures*, Academic Press, New York. Contains a useful short list of further references.

Suppe, J. (1985) *Principles of Structural Geology*, Prentice-Hall, Englewood Cliffs, New Jersey.

11 Emplacement of igneous intrusions

There are four principal methods whereby magmas are emplaced within the crust to form igneous intrusions. The first is a *dilational* process where the country rock moves aside, often under a tensional stress, to allow the magma to fill the space that has been created by the movement (Figure 11.1A). The country rocks are not significantly deformed by this process. The second method, termed *forceful emplacement* (Figure 11.1B), is where the intrusion makes space for itself by forcing the country rocks aside by deforming them. Both these methods are examples of '*active*' emplacement where the pressure of the magma plays a significant part in creating the necessary space for the intrusion.

The third method is *stoping* (Figure 11.1*C*) where the magma moves upwards by removing blocks of country rock which then sink downwards. The fourth method is *melting and assimilation* (Figure 11.1*D*) where the magma makes space for itself by melting and incorporating the country rock. These last two methods are examples of '*passive*' or *permitted* intrusion.

Melting does not appear to be an important mechanism on its own, although it probably plays an important secondary role in many deep-seated intrusions. Stoping as a method of permitted intrusion is an important mechanism in certain high-level plutons but in many other cases probably takes only a minor part in the emplacement process. Neither melting nor stoping is of very great interest to the structural geologist since their roles in deformation are negligible.

Dilation and forceful emplacement on the other hand both have important structural implications. Dilation is the main method of intrusion of the sheet-like bodies and is controlled by the existing stress field. Forceful emplacement is important in large plutonic bodies which produce their own stress field.

11.1 Dilational emplacement of dykes and sills

The stress conditions governing *dyke intrusion* are shown in Figure 11.2*A*. The dyke may be considered to propagate itself upwards by a wedging effect, where the magma pressure (which is hydrostatic) acts perpendicular to the *intrusion plane* of the dyke. In a homogeneous body, the intrusion plane will correspond to the plane of $\sigma_1\sigma_2$ normal to σ_3 and emplacement will take place on condition that:

$$p \geqslant |\sigma_t| + \sigma_3 \qquad (11.1)$$

where p is the magma pressure and σ_t the tensile strength of the wall-rock.

Thus from the orientation of a regional dyke swarm, if we assume the country rock to be structurally homogeneous, we may deduce the orientation of σ_3. If structural heterogeneities exist in the country rock, however, the relationship is not so simple. If a previous fracture plane exists in the rock, emplacement of magma will take place along it, provided only that the magma pressure exceeds the compressive stress across the plane. However, in practice, only fracture planes making a relatively large angle with σ_3 are likely to be chosen.

A dilation

B stoping

C forceful emplacement

D melting and assimilation

Figure 11.1 Mechanisms of igneous intrusion.

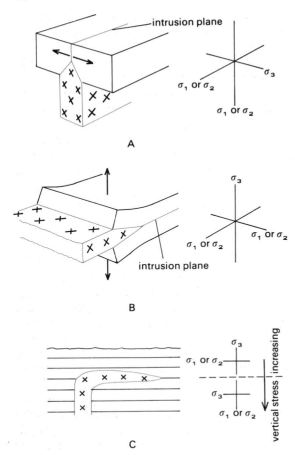

A

B

C

Figure 11.2 Emplacement of dykes and sills. *A.* Dyke emplacement—intrusion plane ideally perpendicular to σ_3. *B.* Sill emplacement—implies a near-vertical orientation of σ_3. *C.* A dyke may feed a sill at a level determined by the change from σ_1 or σ_2 vertical to σ_3 vertical (see text).

Sill formation. Sills represent a special case of sheet emplacement where the intrusion plane is normally sub-horizontal (Figure 11.2*B*). For emplacement to take place, the magma pressure must exceed the load pressure caused by the overlying strata. This will normally occur only at high levels in the crust. We might expect, therefore, that a dyke may become a sill at a level determined by a minimum value of the vertical stress (which increases with depth). If the ratio of the two horizontal principal stresses remains the same, the decrease in the vertical stress will cause the stress axes to swap over, and a dyke may

then change intrusion direction to become a sill (Figure 11.2*C*).

Control by pre-existing structure. A well-known example of the use of faults as well as bedding in the emplacement of a sill is provided by the Stirling Castle sill (Figure 11.3). Here the beds and the faults dip in opposite directions and the sill is able to keep to approximately the same favourable level by following the bedding dip down for a certain distance but periodically returning to a higher level along a fault.

En-echelon emplacement. A rather different kind of structural control is shown by en-echelon emplacement, where a set of intrusions occupies parallel planes which are consistently offset (Figure 11.4*A*). This arrangement may have several different causes. The source of the magma may have been an intrusion in a plane oblique to the individual intrusions but parallel to the zone defined by the set of intrusions (Figure 11.4*B*). This could be explained either by a change in the stress field between two successive levels in the crust, or more likely, by the availability of a set of fracture planes at the higher level oriented obliquely to the intrusion plane at the lower level. A common reason for en-echelon arrangement is intrusion under simple shear stress; thus, for example, a dextral shear stress parallel to the main intrusion plane of Figure 11.4 would tend to open oblique extensional fissures as shown (see Figure 4.2 for a good example of this phenomenon).

Figure 11.3 The influence of pre-existing faults on the emplacement of the Stirling Castle sill in the Midland Valley of Scotland. The sill appears to have used both bedding and faults to maintain an approximately constant level. After McGregor, M.D. and McGregor, A.G. (1948) *British Regional Geology: the Midland Valley of Scotland*, HMSO, London, Fig. 15. TH, Top Hosie Lst., M, Murrayshall Lst., LK, Lower Knott coal.

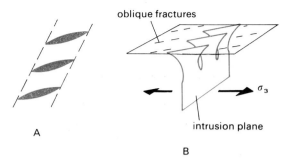

Figure 11.4 Explanation of en-echelon dykes by the filling of oblique fractures. *A*. Plan view of en-echelon dykes. *B*. How the en-echelon dykes may be related to a single intrusion plane at depth.

11.2 Emplacement of cone-sheets and radial dykes

Sets of sheet-like intrusions are commonly associated spatially with high-level plutons in *central igneous complexes* such as the well-documented examples of Ardnamurchan, Mull and Skye (see section 4.3). It is believed that the pattern and orientation of these intrusions are related to the local stress field generated by a pluton situated at depth below the centre of the complex (Figure 11.5). The magma pressure of the pluton exerts a compressive stress perpendicular to the margins of the body resulting in curved sets of stress trajectories.

A possible arrangement of stress trajectories is shown in Figure 11.5A, assuming a simple dome-shaped pluton with circular cross-section. The arrangement has an umbrella shape with σ_1 corresponding to the radially-arranged spokes, perpendicular to the surface of the pluton. The ribs of the umbrella are a set of parabolic curves which may correspond either to σ_2 or to σ_3. The third set of trajectories forms a set of horizontal concentric circles parallel to the rim of the umbrella. This arrangement can explain both the *cone-sheets* and the *radial dykes* that are associated with igneous complexes.

If the circular trajectories correspond to σ_3 then a set of vertical *radial dykes* may form (Figure 11.5D), whereas if the parabolic trajectories correspond to σ_3 then a set of *cone sheets* may form (Figure 11.5B). It is possible that

swapping of σ_2 and σ_3 axes may result from an increase of pressure brought about by the intrusions themselves. This may be the explanation for alternations of cone-sheet and radial dyke intrusion found in some central complexes.

It has been suggested that sets of conical shear fractures generated by this type of stress field (Figure 11.5E) might be used as intrusion planes for both cone sheets and *ring dykes*. The former would follow inward-dipping shear surfaces and the latter outward-dipping surfaces. Such an explanation would relate all three types of minor intrusion found in central igneous complexes to the same basic mechanism.

11.3 Mode of emplacement of large intrusions

From a structural point of view, the most interesting problem associated with the large intrusions is how they became emplaced within the crust and, in particular, how the space they occupy was created—the so-called *'space problem'*. This problem becomes acute when we consider the size of a *batholith*. Has the country rock moved aside to accommodate this vast volume of rock; or has it become, in part at least, digested or assimilated by the magma; or has the country rock been somehow transformed ('granitized') to produce the igneous body? Each of these possibilities has been argued forcefully in the past by its proponents in the 'granite controversy'. We need not concern ourselves with the detailed arguments as to the origin of granites, but there are important structural considerations which bear on the method of emplacement.

In the case of *plutons* emplaced as *permitted intrusions*, the structural relationships suggest a passive accommodation of the intrusive magma to the space left by the country rock as it moves aside, or subsides below an intrusion. Certain large ring-shaped intrusions have been regarded as permitted, created by the subsidence of a central cylindrical block (Figure 11.6). Other plutons appear to have been emplaced by the foundering of blocks of country rock detached from the roof of the intrusion in a form of large-scale stoping (see Figure 11.1C).

Plutons emplaced as *forceful intrusions*, in

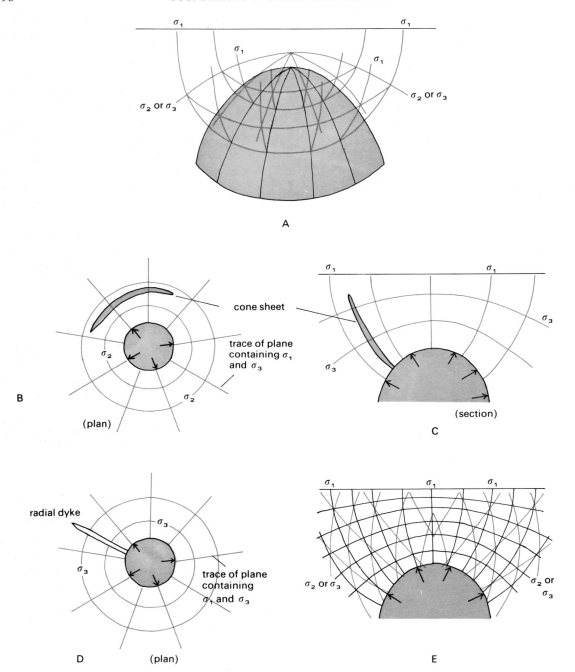

Figure 11.5 Emplacement of cone sheets and radial dykes explained by stress fields generated by a pluton at depth. *A.* Stress trajectories generated by a dome-shaped pluton. Only σ_1 is uniquely determined. *B, C.* Orientation of stresses for the generation of cone sheets (*B*, plan, *C*, section). *D.* Orientation of stresses for the generation of radial dykes (plan view). *E.* Orientation of shear fractures generated by the stress system.

Figure 11.6 Emplacement of a ring dyke by the subsidence of a central cylindrical block.

contrast, make space for themselves by actively pushing aside the surrounding country rock. Evidence of this is provided by folding and fracturing of the strata surrounding the intrusion. A good example of a forceful pluton is the

5 km

Figure 11.7 The Arran granite stock: an example of forceful emplacement. The dotted ornament indicates Upper Devonian and younger strata. Form lines (continuous) and faults (broken lines) indicate the arrangement of folds and faults around the margin of the pluton. After Read, H.H. and Watson, J. (1962) *Introduction to Geology*, Macmillan, London, Fig. 232.

Arran granite stock in south-western Scotland (Figure 11.7). An arcuate fold follows the northern margin of this body, and there are also a number of arcuate faults parallel to the margin. Both the fold and the faults appear to be formed by the forceful emplacement of the granite. Most plutons show some evidence of forceful intrusion.

It seems likely that many of the major granite bodies (whatever the origin of the magma) are *diapirs* (see Figure 4.1*G*) formed as a result of the magma spreading outwards at a particular level of the crust, and do not necessarily extend to great depth. If this is the case, the space problem becomes much less severe than if we were to envisage large steep-sided batholiths extending to the base of the crust. Diapiric emplacement of igneous bodies also takes place in the solid state and is discussed in the following section.

Further reading

Anderson, E.M. (1951) *The Dynamics of Faulting and Dyke Formation with Applications to Britain*, Oliver and Boyd, Edinburgh.

Newall, G. and Rast, N. (eds.) (1970) *Mechanism of Igneous Intrusion*. Geological Journal Special Issue No 2, Seel House Press, Liverpool.

Suppe, J. (1985) *Principles of Structural Geology*, Prentice-Hall, Englewood Cliffs, New Jersey. Contains a useful chapter on intrusive and extrusive structures, with a short list of selected references.

12 Gravity-controlled structures

The force of gravity affects all natural deformation processes and is an important component of all natural stress fields. The role of gravitational load pressure in controlling deformation has already been discussed. However, in this chapter, we deal with structures that are primarily due to the action of gravitational force rather than to an applied external stress.

12.1 The effect of topographic relief

Differential erosion commonly leads to gravitational instability by exposing the ends of a set of beds along valley sides (Figure 12.1). The effects range from comparatively minor bending

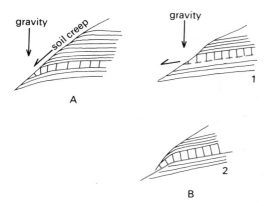

Figure 12.1 Structures formed as a result of gravitational instability associated with a topographic slope. *A*. Bending of strata associated with soil creep. *B*. Slip of a competent layer over a weaker layer.

of the strata close the the ground surface, associated with *soil creep*, to *gravitational collapse structures* many hundreds of metres in length. The instability produced by a topographic slope is greatly accentuated if the beds dip towards the slope (Figure 12.1*B*). Competent layers resting on weak material may then slip towards the valley under gravity, particularly if their slip plane is weakened or lubricated by percolating ground water.

Several types of structure attributed to gravitational collapse have been described from the mountains of Iran, where a series of thick compe-

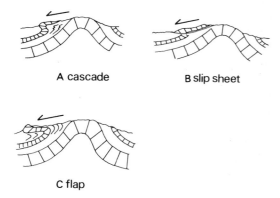

Figure 12.2 Examples of gravity collapse structures from Iran. After Harrison, J.V. and Falcon, N.L., *Quart. J. geol. Soc. Lond.* (1936) **92**, 91–102. *A*. Cascade. *B*. Slip sheet. *C*. Flap (see text for explanation).

tent limestones overlie incompetent anhydrite-bearing shales. The structures include *recumbent folds* and *slip sheets* resting on faults (Figure 12.2) and are attributed to the instability of dipping competent sheets on the limbs of anticlinal ridges. The instability may be accentuated by the effect of erosion on the crests of the ridges.

12.2 Gravity sliding of sheets or nappes

A *nappe* is a sheet of rock, of the order of tens of kilometres in extent, that has been displaced along a basal fault plane. Many nappes have a thrust base and are compressional in origin, but others have been attributed to gravitational sliding along a low-angle normal fault or *lag*. Because of subsequent changes in the attitude of the fault planes, it is not always possible to be certain whether a particular fault plane is a thrust or a plane of gravity sliding.

In the past there has been considerable controversy over the origin of the great nappes of the French and Swiss Alps, one school holding that the nappes were formed by gravitational gliding from an uplifted crystalline mass in the interior, and another claiming that the nappes were basically compressional in origin and formed by upwards and outwards flow of material squeezed from a highly compressed central root zone. Some Alpine geologists believe that both mechanisms play a part in the overall structure. An initial compression produces an uplift, which is followed by gravitational gliding down the slopes of the uplift.

A true *gravity-gliding nappe* or *sheet* should exhibit an *extensional* zone at the upper or *proximal end* of the sheet (i.e. nearest the uplift) and a *compressional* zone at the lower or *distal end* (i.e. furthest from the uplift). Low-angle normal faults on which the sheets rest are termed *detachment faults*. These faults cut up to the surface at the upper end of the sheet to allow the sheet to become detached. The Bearpaw Mountains of Montana and Wyoming, near the eastern edge of the Cordilleran orogenic belt of the western USA, are an elevated block, 50 km across, of Eocene volcanic rocks surrounded by horizontal sedimentary strata that are folded and

Figure 12.3 Cross-section of gravity-sliding structure on the north flank of the Bearpaw Mountains, Montana (see text). From Suppe, J. (1985) *Principles of Structural Geology*, Prentice-Hall, New York, Fig. 8.30, with permission.

faulted adjacent to the block. The sense of normal slip on the faults indicates movement away from an uplifted Tertiary intrusive complex at the southern edge of the range (see Figure 12.3). Drill hole data indicate that the volcanics rest on a gently inclined detachment horizon in weak bentonitic shales of Cretaceous age. The only reasonable interpretation of this structure is that the Bearpaw Mountains sheet has slid on a very gently inclined décollement surface off the uplifted laccolithic intrusive complex. As pointed out previously (section 9.1) a high pore-fluid pressure will reduce the angle necessary for sliding to a few degrees.

12.3 Salt domes and diapirs

Salt domes arise from the gravitational instability of a layer of low density (salt, or other evaporite material) overlain by rock of higher density. If the salt layer and the strata above are perfectly regular and uniform there is no tendency for the salt to move. However, any heterogeneity in the system caused, for example, by a thickening of the salt layer, by folding or faulting in the beds above, or by local erosion, will lead to lateral pressure in the salt layer induced by the gravitational load. This pressure would lead ultimately, if able to continue to completion, to the flow of all the salt to the surface, where it would form a layer in a new gravitationally stable position.

In practice of course, the above process is only partly completed. Structures caused by the movement of salt show a wide variety of forms reflecting different stages in its upward migration, commencing with simple broad domes and proceeding to plug-like and mushroom-shaped forms (Figures 12.4, 12.5). Such structures are termed *diapirs* and the process *diapirism*.

Structures associated with salt diapirs. All salt diapirs show marked upwards bending of the surrounding layers against the walls of the body, often accompanied by reverse faulting. In certain cases tight folding may occur in ductile layers. Thus the essentially cylindrical body of the salt intrusion is surrounded by a ring-shaped zone of compressional folding and faulting (Figure 12.4).

The strata above the salt body are affected by extensional tectonics resulting in arching and thinning of layers and in normal faulting. Intricate patterns of normal faults are commonly found above salt domes.

Within the salt body, the strain pattern is dictated by the direction and extent of flow. In a structure like that shown in Figure 12.4 (5), the narrow 'neck' will exhibit very tight folding with a strongly-developed vertical linear elongation.

12.4 Mantled gneiss domes

The crystalline regions in the cores of many orogenic belts and in granitic or gneissose Pre-

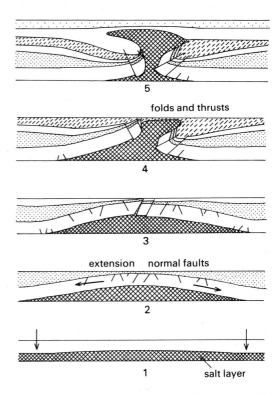

Figure 12.4 Evolution of a salt diapir arising from the initial gravitational instability of a salt layer overlain by denser strata. As the diapir evolves, successive layers are deposited on the surface, each in turn becoming deformed as the diapir migrates upwards. Note the early extensional structures associated with the doming and later compressional structures associated with the 'neck' of the diapir. After Trusheim, F. (1960) *Bull. Am. Ass. Petrol. Geol.* **44**, 1519–1541.

Figure 12.5 Forms of salt structures in the Hanoverian salt-dome region of N Germany. A Permian salt layer (coloured) overlying Palaeozoic basement is overlain in turn by Triassic to Cretaceous strata which are penetrated and deformed by the salt bodies. After De Sitter, L.U. (1964) *Structural Geology*, McGraw-Hill, New York, Fig. 162.

cambrian shields commonly exhibit dome-shaped areas of granitic material surrounded by a 'mantle' of metasedimentary or metavolcanic rocks. Structures of this type were first described from Finland and were given the name '*mantled gneiss domes*'. In many cases the basal mantle rock is a conglomerate, which indicates that the dome structure is a deformed unconformity between cover and basement. The material of the dome itself is often a complex mixture of rock types which, although predominantly granitic in composition, may include bands or small bodies of many other rock types. Some domes consist entirely of granite which might suggest that the domes are intrusive bodies. However, this explanation fails to account for the mantling effect and is a feasible origin only for domes with discordant margins.

It is believed by many geologists that most gneiss domes have originated in the same way as the salt domes described in the previous section; that is, by the gravitational instability of a lower-density granitic layer overlain by a denser layer of, for example, basic volcanics. The mechanism has been tested experimentally using a centrifuge, and very similar structures have been produced (Figure 12.6). Other geologists have suggested that many domes may be the result of interference of two fold sets (see Figure 10.18*A*) producing domes at the culminations of crossing anticlines. However, folding would tend not to produce equidimensional patterns everywhere and we might expect there to be some evidence of such large-scale folding in adjoining areas. Careful mapping of structures within certain domes has revealed patterns distinctive of a diapiric process and is the best method of establishing the origin.

12.5 Large-scale structures

Depressed or elevated areas representing large-scale warps or flexures with wavelengths of the

Figure 12.6 Model of mantled gneiss dome structure produced in a centrifuge experiment. Domes of stitching wax in putty overburden are developed along the boundary of a deep-seated less dense source layer. Diameter of model 94 mm. The lower photograph shows a sideways view with the putty overburden removed. The domes are outlined in ink. This experiment simulates the effect of gravity in causing the diapiric uprise of less dense material into a denser overburden. From Ramberg, H. (1962) *Bull. geol. Inst. Univ. Uppsala* **42**, Figs. 27, 28.

Figure 12.7 Map showing the shape of the Michigan basin, Northern USA. Contours are drawn on the position of the Coldwater formation at depth, at intervals of 500 feet. Based on De Sitter, L.U. (1964) *Structural Geology*, McGraw-Hill, New York, Fig. 299.

order of hundreds of kilometres upwards are very characteristic features of the Earth's crust. Structures of this type are particularly important in the 'stable' non-orogenic areas or *cratons* where they are usually referred to as *basins* and *uplifts* respectively (see section 13.4).

The main Mesozoic–Tertiary basins of northwestern Europe, for example, have a typically equidimensional shape with no pronounced elongation or alignment. The intervening areas are more irregular in shape and represent the uplifts which have contributed the sedimentary fill to these basins over this period. Most basins evolve gradually by the successive addition of formations, many of which may thicken towards the centre of the basin and thin towards its

margins, the cumulative effect being to enhance the basin shape of the floor. Compared to compressional fold structures, the dips are very low—of the order of 1° or less. Figure 12.7 shows the shape of the Michigan basin in the northern USA by means of a set of contours drawn at successively deeper levels of the Coldwater Formation. These show that the basin has an approximately circular plan, about 380 km across and over 750 km deep.

Such structures are considered to be primarily gravitational in origin. Horizontal tectonic compression can be ruled out by the large size and lack of orientation of the structures. The principle of gravitational equilibrium (*isostasy*) requires any large mass deficiency in the crust, such as a basin, to be compensated by a mass of greater density beneath, so as to make the weight of that column through the Earth balance that of the adjoining columns. This compensation could be achieved by a thinner crust, so that the denser upper mantle material was closer to the surface—analogous to, but on a lesser scale than, the ocean basins. Some large basins are thought to originate by extensional thinning, whereas others may be due to density variations within the upper mantle. Uplifts may be formed by the reverse process.

When any disturbance of gravitational equilibrium takes place, gravitational forces act to restore that equilibrium by flow of material at depth and compensating uplift or depression at the Earth's surface.

Further reading

Ramberg, H. (1967) *Gravity, Deformation and the Earth's Crust*, Academic Press, London. Experimental evidence for gravitational structures.

Schwerdtner, W.M., Stone, D., Osadetz, K., Morgan, J. and Stott, G.M. (1979) Granitoid complexes and the Archaean tectonic record in the southern part of NW Ontario. *Can. J. Earth Sci.* **16**, 1965–1977. Discusses the formation of mantled gneiss domes by diapiric emplacement in the solid state.

Part 3: Geotectonics—major Earth structure

In this section of the book, we discuss the significance of geological structures in the context of large-scale Earth processes. The development over the last two decades of plate tectonic theories gives us a way of integrating rock deformation into a model that attempts to explain the evolution of the crust as a whole. As a background to the plate tectonic hypothesis, which is dealt with in section 14, we discuss the major Earth structures at the present day and, in particular, the distribution of tectonic activity. Section 15 relates deformation to the plate tectonic model and examines the nature of this relationship in the geological past back into the Precambrian.

13 Major Earth structure

If we can imagine the Earth viewed from space, with the water of the oceans removed, the major features of the structure of the crust would immediately be apparent (Figure 13.1). A first-order division of the crust can be made into *continents* and *oceans*, and superimposed on these large features are rather linear elevations consisting of the *mountain ranges* on the continents and the great system of *ridges* in the oceans. Of a smaller order of magnitude are the deep ocean *trenches*. In this chapter, we shall examine the nature and pattern of these large-scale structures, and also their relationship to the tectonically active zones of the crust.

13.1 Continents and oceans

The distribution of topographic level of the Earth's surface is shown in Figure 13.2. It is clear that there are two dominant levels corresponding to the continents (average height about 1 km) and the ocean basins (average depth about 4 km)

Figure 13.1 The major topographic features of the Earth's surface as they would appear with the ocean waters removed. After Wyllie (1976), Fig. 3.7.

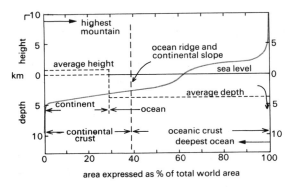

Figure 13.2 Distribution of topographic level of the Earth's surface expressed as a proportion of total surface area (see text). After Wyllie (1976), Fig. 3.11.

Figure 13.3 Average cross-sections through continental and ocean crust showing thicknesses and mean densities.

respectively, and that the proportion of the total area occupied by the extremes of height and depth (mountain ranges and ocean trenches) is very small. Elevations greater than 3 km make up only 1.6%, and depressions deeper than 5 km only 1%, of the total area.

The continents cover 29% of the Earth's surface, distributed in a rather uneven way with 65% of the total land area in the northern hemisphere. The distribution of land and sea has varied very considerably through geological time, and there has been a continuous change both of the relative positions of the continents and of the shoreline position. If we add the continental shelf and slope to the area of the continents, the total continental surface area is 40%, compared with 60% for the ocean basins, which more exactly reflects the relative proportions of continental to oceanic crust.

The existence of such a large difference in level is explained primarily by the difference in thickness between continental and oceanic crust. The principle of general gravitational balance (*isostasy*) means that topographically higher sections must contain a greater proportion of lower density material to keep the total weight the same (see section 12.5). The base of the crust, at the *Mohorovičić discontinuity* (*Moho*), marks a very significant change in composition and density (Figure 13.3). The mean density of crustal rocks is around 2.8–2.9, whereas the peridotitic rocks of the uppermost mantle have a mean density of

around 3.4. Continents are situated on crust with an average thickness of 33 km and a mean composition close to that of granite, whereas ocean basins are situated on crust with an average thickness of only 7 km and formed mainly of rocks with a gabbroic or basaltic composition (Figure 13.3). The ability of rocks to flow at depth means that material within the mantle can be transferred so as to maintain isostatic equilibrium and allow each part of the crust to sink or rise to the appropriate level. The difference of over 4 km in mean elevation between continents and oceans is therefore explained by the buoyancy of the thick continental crust.

13.2 Mountain ranges, ocean ridges and trenches

The elevated and depressed features of the continents and oceans are generally linear and more or less continuous over long distances.

Mountain ranges. The high mountain ranges of the Earth form two main belts, one situated on the western side of North and South America, and the other (the 'Alpine–Himalayan belt') approximately at right angles to it, forming a sinuous belt from the Mediterranean through central Asia, and curving southwards through Burma to Indonesia. Another type of mountain belt forms the series of *island arcs* which are found around the northern and western margins of the Pacific Ocean and in the north-eastern Indian Ocean. The topographic relief of these chains is comparable to that of the continental ranges, but because they are partly submerged they appear less significant. These elevated features show extreme variation in dimensions but as a very rough approximation we can regard their width as generally 300–800 km, and their ratio of vertical to horizontal dimensions as in the range 1:100–1:200.

Ocean ridges. The great network of ocean ridges which figure so prominently on the Earth's surface (Figure 13.1) represent topographic relief of a rather greater order of magnitude volumetrically than the continental mountain ranges. They occupy about one-third of the surface area of the oceans and rise to between 2 km and 3 km from the ocean floor. Their width varies but is typically in the range 500–1000 km, and the ratio of their vertical to horizontal dimensions is thus around 1:500. These huge structures are isostatically compensated, the excess topographic relief being balanced by hotter, less dense mantle material beneath.

Ocean trenches. The deep ocean trenches form a discontinuous system of arcuate features which either lie near the continental margins (as in the South American trench) or border island arcs as in the north and west Pacific. They differ in

dimensions from both the other types of linear feature, being generally around 100–150 km in width, and 2–3 km deep. The deepest trenches are over 11 km below sea level. They are thus very narrow deep features with a depth/width ratio of around 1:50.

13.3 Present-day tectonic activity

The pattern of current tectonic activity is very closely related to the topographic pattern we have just described. This activity is of three types: (i) *seismic movements*—mainly displacements of the crust with high strain rates, (ii) *aseismic* crustal movements with low strain rates, and (iii) *vulcanicity*.

Seismicity. The distribution of seismicity is shown in Figure 13.4. The vast majority of earthquakes, including all the severe ones, are concentrated in narrow belts which correspond with the linear topographic anomalies. More than 80% of the total earthquake energy is concentrated in the circum-Pacific belt alone. Thus the young mountain ranges, ocean ridges

Figure 13.4 Main belts of recent earthquake activity. Solid colour, shallow earthquakes only (< 65 km); vertical ruling, mainly shallow to intermediate (65–300 km); stipple, shallow to deep (> 300 km). Based on Chadwick, P. (1962) in *Continental Drift*, ed. S.K. Runcorn, Academic Press, New York, Fig. 4, and Toksoz, M.N. (1975) *Scientific American*, November.

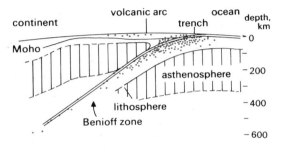

Figure 13.5 Section across the Japanese island are showing the concentration of seismic activity along the Benioff zone. Earthquake foci are shown as coloured dots. After Uyeda (1971), Fig. 5.18.

and ocean trenches, which represent extreme disturbances of the Earth's relief, are also the sites of severe tectonic activity. If we look at the distribution in terms of the depth of *focus* or point of origin of the earthquakes, we find that the ocean ridges exhibit only shallow earthquakes, with focal depths down to 65 km, and that these are concentrated along a central rift zone or along faults which offset that zone. The deep earthquakes, with focal depths of over 300 km, are concentrated along the deep ocean trenches,

especially around the Pacific. A typical cross-section of a tectonically active segment of the northern or western Pacific shows that the earthquake foci lie on a plane outcropping in a trench and dipping below an adjacent island arc or continental margin. This inclined zone of earthquake activity is called a *Benioff zone* (Figure 13.5) and is a critical piece of evidence in favour of the process of *subduction* (see section 14.3).

Focal plane solutions (see section 9.3) generally indicate a compressive component acting across the Benioff zones, whereas ocean ridge earthquakes show tensional solutions. The great oceanic faults (e.g. Figure 14.15) are generally strike-slip.

Aseismic movements. Slow movements of the crust of the order of millimetres per year are detectable by precise measurements of height over a period of years. These rates of movement are comparable with the rates determined from the geological evidence of uplift and sedimentation. All parts of the crust are subject to these aseismic movements which are mainly vertical.

Figure 13.6 World distribution of active volcanoes. Currently active volcanic areas are shown in solid colour. The seismic zones of Fig. 13.5 are shown for comparison.

Vulcanicity. The distribution of present-day vulcanicity (Figure 13.6) bears a striking resemblance to the earthquake distribution (Figure 13.4). Clearly there is a very close relationship between vulcanicity and zones of tectonic instability. About 75% of the currently or historically active volcanoes are situated in the circum-Pacific belt, particularly along the *volcanic island arcs*. Many volcanoes are associated with the ocean ridge network but some are also found along faults or lineaments within the ocean basins. The Alpine–Himalayan belt exhibits a rather sparse distribution of volcanoes, with the exception of the Mediterranean and Indonesian regions where there is adjoining oceanic crust with sections of trench. The greatest concentration of vulcanicity within the continents is found in the African rift system which is a continental extension of the Indian Ocean ridge network. ·

13.4 Stable and unstable tectonic zones

The clear subdivision which can be made at present into tectonically *stable* and tectonically *active* or *unstable* zones is one that characterizes most of the geological record. We have no direct information on the nature of the oceans before about 200 Ma ago, but the continental areas for any particular period, extending back to about the mid-Precambrian (*c.* 2500 Ma ago) show a contrast between stable zones called *cratons* and unstable zones called *mobile belts* or *orogenic belts*.

The *cratons* consist of undeformed flat-lying sedimentary cover on an older crystalline basement and exhibit only very minor vulcanicity. Tectonic effects are confined to very slow vertical tectonic movements of the order of millimetres per year or less, resulting in broad sedimentary basins separated by uplifts (see section 12.5). The orogenic belts contain highly deformed sediments, abundant and varied igneous rocks, and uplifted segments of deep crustal material. The tectonic processes which take place in these mobile belts generally result in the formation of mountain belts. This is a result of crustal thickening, partly compressional and partly volcanic in

origin, which must lead ultimately to uplift due to isostatic forces.

The continental crust of the present-day stable regions is composed of the worn-down remnants of previous orogenic belts of different ages. The proportion of orogenic belt to craton appears to increase generally as we go further back in time, until in the early Precambrian no trace of cratons can be found and mobility appears to be general.

Further reading

See section 14 references.

14 Plate tectonics

14.1 Historical context

In 1915, Alfred Wegener popularized the idea of *continental drift* in his book *The Origin of Continents and Oceans*. The relative movement of continents provided a convenient explanation for many otherwise puzzling geological phenomena—the geometric fit of the opposing coastlines of the Atlantic, peculiarities in faunal distribution and palaeo-climatological reconstructions, and awkward geological coincidences, that immediately made sense when continents were joined together in their presumed original positions to form a supercontinent, 'Pangaea'.

Many geologists found difficulty in accepting the proposed movements in the absence of a satisfactory mechanism, but many of the doubters were convinced by the results of palaeomagnetic work in the period 1950–60. Palaeomagnetic reconstructions for the Triassic period showed that the continents, when reassembled in the positions advocated by the continental drifters, showed a common Triassic magnetic pole (Figure 14.1*A*). The positions of the Triassic poles for the various continents at present are widely scattered. An even more convincing test of the reassembly is provided by the *apparent polar wander curves*. These tracks of successive positions of the magnetic north pole are shown for the continents of Europe and North America from the Cambrian to the present (Figure 14.1*B*).

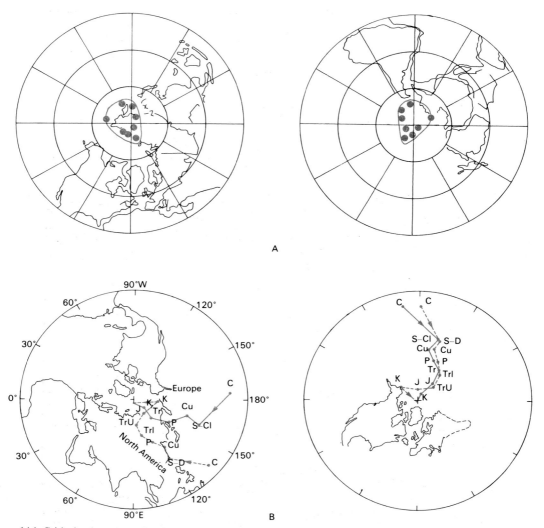

Figure 14.1 Critical palaeomagnetic evidence for continental drift. *A.* North and south polar projections of the continents re-assembled in their pre-drift positions, showing the positions of the Triassic magnetic pole for the various continents (coloured dots). *B.* The left-hand diagram shows the apparent polar wander curves for Europe and North America in their present relative positions. The right-hand diagram shows the superimposition of the curves between Silurian and Upper Triassic times with the continents in their pre-drift positions. C, Cambrian; S, Silurian; D, Devonian; Cl, Lower Carboniferous; Cu, Upper Carboniferous; P, Permian; Tr, Triassic; Trl, Lower Triassic; Tru, Upper Triassic; J, Jurassic; K, Cretaceous. After McElhinny, N.W. (1973) *Palaeomagnetism and Plate Tectonics*, Cambridge University Press, Cambridge.

The tracks for the Carboniferous to Jurassic period, when these continents were thought to be joined together, are precisely superimposed when the continents are reassembled but diverge for the period after the Jurassic when opening of the North Atlantic commenced. The apparent polar wander curves are of course quite separate in the present positions of the continents.

The acceptance by most geologists of the idea of continental drift led to attempts to discover a mechanism, and in particular to understand the role of the oceanic areas in this process. The hypothesis of *sea-floor spreading*, put forward in 1962 by Hess and Dietz, linked continental and oceanic areas in the same general process, likened to a conveyor belt, in which slabs of crust were

Figure 14.2 Magnetic stratigraphy of the ocean floors. *A*. The magnetic 'tape recorder'. Diagrammatic representation of the process whereby sea-floor spreading and magnetic polarity reversals produce a series of differently magnetized stripes parallel to the ocean ridge crest. (*a*) at 2.75 Ma, (*b*) at 2.25 Ma, (*c*) at present. From Wyllie, (1976), Fig. 10.5, with permission.

Figure 14.2 (*contd.*) Magnetic stratigraphy of the ocean floors. *B.* Time scale for the last 4 Ma.

transported laterally carrying the continents with them. The *ocean ridges* were envisaged as the sites of *creation* of new oceanic material and the *trenches* as the sites of *destruction* (see Figures 13.5 and 14.7).

Thus the two hypotheses of continental drift and sea-floor spreading became combined to form the nucleus of the new hypothesis of *plate tectonics* which has evolved into a general theory to explain crustal movements and evolution. The critical evidence that led to the acceptance of the sea-floor spreading idea arose from the application of *magnetic stratigraphy* to the ocean floor, from which the direction and amount of movement of the ocean floor could be established.

Measurement of the magnetic field over the oceans had shown that the ocean floor exhibited a striped magnetic pattern with abrupt changes in magnetic intensity from one stripe to the next (Figure 14.2). These stripes, which are mostly around 20–30 km in width, are parallel to the ocean ridges and, like them, are displaced by the oceanic fracture zones. A great breakthrough came in 1963 with the proposal that these abrupt changes in magnetization could be correlated with *reversals of polarity* in the Earth's magnetic field. Since these reversals could be dated by comparison with continental rock sequences (Figure 14.2*B*), the magnetic stripe pattern effectively became a magnetic stratigraphy of the ocean floor, from which it is possible to establish

the age of a large proportion of the oceanic crust (e.g. see Figure 14.8) and thus to prove that the ocean spreading hypothesis was correct.

14.2 The concept of lithosphere plates

The idea of *plates* arose from the observation that large areas of the crust have suffered very little distortion although they have travelled laterally several thousand kilometres, if we accept the evidence for continental drift. The detailed and accurate 'jigsaw' fit of the opposing coastline of America and Africa (Figure 14.3), after 200 Ma and 4000 km of drift, testifies to this lack of distortion. In the oceans also, we find regular linear magnetic stripes and faults which have maintained their shape after tens of millions of years. This evidence reinforces the conclusions reached by studying the distribution of tectonic movements (see section 13.4) that there are large stable areas (the *cratons*) that suffer little internal deformation and exhibit only slow vertical movements, while moving laterally as a coherent unit at rates 10–100 times faster.

Figure 14.3 Geometric fit of the opposing continental margins of the Atlantic. Solid colour represents areas of misfit. Matched at 1000 m below sea level. After Bullard, E.C., Everett, J.E. and Smith, A.G. (1965) *Phil. Trans. R. Soc.* **258A**, 41–51.

Seismicity and plate boundaries. The obvious link between seismicity and present-day tectonic activity suggested that the seismic zones must represent the boundaries of these stable blocks of crust and that each block or *plate* could be delimited by a continuous belt of seismic activity. Taking the argument one step further, since the seismic activity represents fault movements with high strain rates, each plate must be in a state of relative motion with respect to each of its neighbours. If we now re-examine the nature of these plate boundaries and their sense of movement, we can recognize three types of *plate boundary* (Figure 14.4)—(i) boundaries where adjoining plates are moving apart (ocean ridges); (ii) boundaries where adjoining plates are moving together (ocean trenches and young mountain belts); and (iii) boundaries where adjoining plates are moving laterally past each other with a horizontal strike-slip sense of displacement along steep faults. The sense of displacement can be deduced from first-motion studies of individual earthquakes and in general confirms the relative movements inferred from other evidence such as palaeomagnetism and magnetic stratigraphy.

Following these principles, we can use the network of boundaries to divide the present Earth's surface into six major plates (Figure 14.5), the *Eurasian, American, African, Indo-Australian, Antarctic* and *Pacific plates.* There are also a number of smaller plates associated with destructive boundaries, especially

Figure 14.5 The six major lithosphere plates. Ridges, double lines; trenches, single lines; transform faults, broken lines. After Le Pichon, X. (1968) *J. geophys. Res.* **73**, 3661–3697.

around the margins of the Pacific Ocean. Note that plate boundaries may or may not correspond to continental margins. Continental margins that lie within plates, such as the Atlantic margins of Africa and America, are termed *passive margins*. Those that do correspond to plate boundaries (e.g. the western margin of the American continents) are termed *active margins*.

Lithosphere and asthenosphere. The vertical extent of a plate is much more difficult to determine than its horizontal extent. Alfred Wegener's original idea of pieces of continental crust moving across a plastic ocean was abandoned many years ago when it was realized that oceanic rocks could not behave in a sufficiently ductile manner near the surface. The possibility that the plates consist of pieces of crust sliding over the mantle must also be discarded. The oceanic crust is only about 7 km thick and could not remain undistorted or transmit horizontal stresses. Moreover, there is no evidence of a major change in physical properties which would suggest the existence of a zone of sub-horizontal displacement at the base of the crust.

The evidence from earthquake waves suggests that the physical properties of the crust and uppermost mantle change gradually down to around 100–150 km in depth, where there is a more abrupt change in the velocity profile. The rate of increase in seismic velocity drops through a zone about 100 km deep before rising again at greater depths. This layer of abnormally low seismic velocities is called the *low-velocity zone* (LVZ) and is thought to signify a decrease in

Figure 14.4 Block diagram illustrating the plate tectonic model. After Isacks, B., Oliver, J. and Sykes, L.R. (1968) *J. geophys. Res.* **73**, 5855–5899.

Figure 14.6 Lithosphere and asthenosphere. The lithosphere includes the crust and the uppermost part of the upper mantle.

density and in viscosity. We can therefore regard this zone as a more ductile layer where lateral flow of material could take place, or as a zone of

ductile shear between plate movements above and the main part of the mantle beneath.

Isostasy theory demanded a weak layer where lateral flow could take place, and this layer was called the *asthenosphere*. It has become convenient to identify the asthenosphere with the LVZ. The stronger layer above the asthenosphere is termed the *lithosphere* (Figure 14.6). The lithosphere thus includes the crust and the uppermost part of the upper mantle down to a variable depth of 80–120 km in the oceans and possibly around 150 km in the continents. Since the base

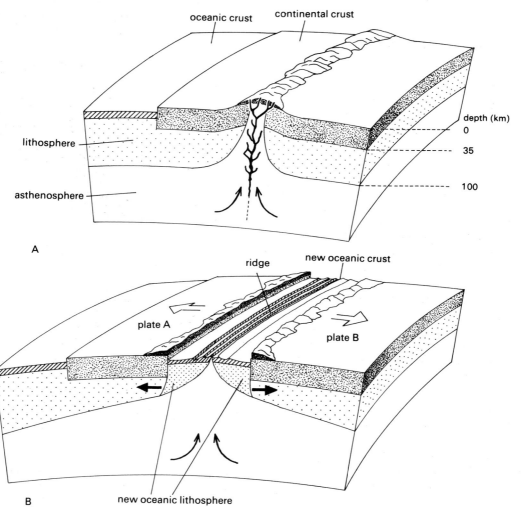

Figure 14.7 Nature of constructive plate boundaries. *A*. Initiation of a constructive boundary as a continental swell due to the uprise of hot asthenosphere. *B*. Formation of new oceanic lithosphere along an ocean ridge between the two diverging plates. After Dietz, R.S. and Holden, J.C. (1970) *Scientific American*, October.

of the lithosphere depends on a relatively gradual change in *viscosity*, which is strongly temperature-dependent, the base is not only gradational but varies both spatially and temporally in response to changes in temperature gradient. If we take as an example the oceanic lithosphere, this will be hottest near the site of formation at the ocean ridge crest and will cool gradually with time and distance from the ridge as it travels laterally away from it. This is reflected in the thickness of the lithosphere which varies from less than 50 km at the ridge crest to 120 km near the ocean margins (Figure 14.7).

14.3 Nature of plate boundaries

Constructive boundaries. Plate boundaries along ocean ridges where adjoining plates are moving apart are called *constructive boundaries*, because new material is being added to them. The process is illustrated in Figure 14.7. Some of the new material is provided by upper mantle melts formed in the hot, low density region below the ridge. Part of this molten material is injected into the crust as basalt dykes or gabbro intrusions, and part is extruded on the ocean floor as basalt pillow lavas. At the same time, the mantle part of the lithosphere grows by the addition of ultrabasic intrusions and by ductile flow of material from the asthenosphere. The focus of activity at any given time is the central seismically-active rift zone marked by dyke intrusion, extensional faulting and vulcanicity. This zone is only about 100 km wide, and the remainder of the ridge consists of warm lithosphere which gradually subsides to the level of the ocean basins as it cools and moves away from the central rift.

The *continental rift zones* represent incipient constructive boundaries. The African rift zone, the Red Sea rift and the Gulf of Aden rift meet in a triple junction in the Afar region of Ethiopia (see Figure 15.1) which serves as a useful analogy for the way in which the major continents separated during the Triassic to Jurassic period. This example will be discussed in more detail in section 15.2. The characteristic features of the oceanic constructive boundaries are also found in the continental rifts but the nature of the vulcanicity is much more varied.

Destructive boundaries. There are two types of convergent boundary at the present time. The first follows the deep *ocean trenches* and the second follows the belt of *young mountain ranges* of the Alpine–Himalayan chain (Figure 14.5). The evidence for the nature of the convergent movement along the trenches comes partly from earthquake first-motion study, which shows generally compressional solutions across the trench, and partly from the shape of the Benioff zone, which suggests that the oceanic plate dips down

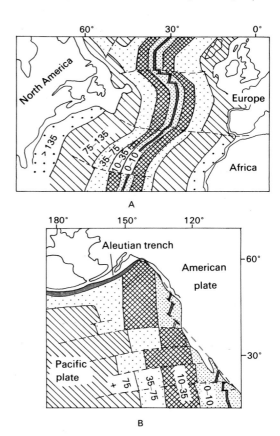

Figure 14.8 Comparison of ocean-floor magnetic stratigraphy at constructive and destructive boundaries. *A*. Magnetic age pattern of the central Atlantic. Note concordance with continental margins. *B*. Magnetic age pattern of the NE Pacific. Note discordance with the continental margin. After Larson, R.L. and Pitman, W.C. (1972) *Bull. geol. Soc. Am.* 3645–3661. (Age in Ma).

below the adjoining plate (see Figures 13.5, 14.10). This process is known as *subduction*, and since subduction involves the destruction of plates, by returning old lithosphere material to the mantle, *subduction zones* are termed *destructive plate boundaries*.

Geological evidence for the subduction process is provided by examining the magnetic stratigraphy of the ocean floor adjacent to a destructive boundary and comparing it with the pattern at a constructive boundary (Figure 14.8). The magnetic stripes in the Atlantic (Figure 14.8*A*) are concordant with the coast lines; the oldest stripes adjoin the continents and reflect the time of separation. In the northern Pacific, in contrast (Figure 14.8*B*), the stripes are discordant with the Aleutian trench and stripes of various ages occur along this plate boundary, demonstrating that their continuations have been subducted below the trench.

Certain subduction zones border continents, on the west side of South America for example. In this case there is a linear volcanic belt situated on the continent about 300 km from the Peru–Chile trench (see Figures 13.1, 13.6). Most subduction zones at the present time, however, are situated at *island arcs* within the oceans (Figure 14.15), so that a section of oceanic crust intervenes between the subduction zone and the nearest continent. A typical island arc (see Figure 13.5) consists of a partially submerged volcanic mountain range

50–100 km wide with a trench on its convex side between 50 and 250 km from the island arc. Certain island arcs are formed from pieces of continental crust that have perhaps become separated from a nearby continent. Others are built up by the addition of new volcanic material to oceanic crust.

The arc-trench zone is not in isostatic balance. There is a mass deficiency along the trench and a smaller mass excess associated with the volcanic arc. This gravitational instability must be related to the subduction process, and the mass imbalance is thought to be supported by the lateral compressive stress associated with the convergent plates. The areas of oceanic crust between the island arc and the nearest continent form what are known as *marginal basins*. These structures have been attributed in many cases to secondary ocean-floor spreading behind the island arc. Figure 14.9 illustrates two possible mechanisms for producing the secondary spreading—the mantle diapir model and the trench (or slab) roll-block model.

The other type of convergent plate boundary is often referred to as a *continental collision zone*. In the case of the Alpine–Himalayan belt, the plate boundary represents the collision of the Eurasian continent to the north with the African and Indian continents to the south (see Figure 14.5). The collision can be deduced from the record of relative movements of these continents, which is

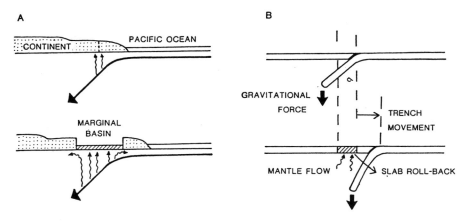

Figure 14.9 Possible mechanisms of formation of a marginal basin. *A*. Secondary spreading due to heating of the upper mantle above a subducting slab. *B*. The trench (or slab) roll-block model. See text. *A*. After Uyeda (1978), Fig. 5.22.

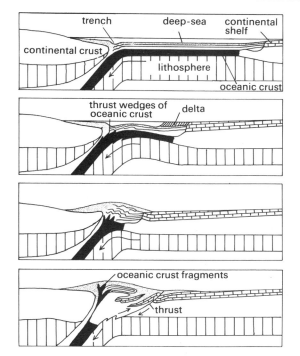

Figure 14.10 Sequence of stages in the transformation of a subduction zone to a continental collision zone by the approach of two continents and the elimination of the intervening ocean. After Dewey, J.F. and Bird, J.M. (1970).

well documented by palaeomagnetic data and ocean-floor stratigraphy. A plate margin of this type is transient in nature since continental crust, because of its low density, is not subductable. Thus convergence of two pieces of continental lithosphere can only take place to a limited extent before the two continents lock together.

The line of collision is often termed a *suture* and is important in recording older plate movements in the geological record of the continents (see section 15.5). Along the Himalayan part of the suture, there is geological evidence of a subduction zone in the recent geological past, which was responsible for destroying a large area of oceanic plate formerly separating Asia and India. The processes leading from a subduction zone to a collision zone are summarized in Figure 14.10.

In the case of the India–Asia collision zone, the convergent plate movements have resulted in a very wide belt of complex structures on the Asian

side of the suture (see section 15.5 and Figure 15.5).

Conservative boundaries. Many sections of the boundaries of all the plates consist of steep faults with a lateral (*strike-slip*) sense of displacement. Because plate material is neither created nor destroyed along these sections but is conserved, they are termed *conservative boundaries.* Faults or fracture zones are very prominent features of the oceans, and the ocean ridge crests are repeatedly offset by them (see Figure 14.15). These oceanic faults played a key role in the evolution of the plate concept. It was noted by J. Tuzo Wilson, in an influential paper in 1965, that parallel sets of such faults should be parallel to the spreading direction of the ocean ridges, and that the divergent motion away from a ridge axis would be 'transformed' to a transcurrent motion along such a fault (Figure 14.11*A*), then perhaps transformed again to convergent motion at a trench (Figure 14.11*B*). He therefore called these faults *transform faults*, recognizing their fundamental difference from strike-slip faults on land. A transform fault is part of a plate boundary, and must be parallel to the direction of relative motion of the plates on either side. It is therefore controlled by the relative velocity of the two plates, whereas a strike-slip fault, at least initially, is a response to stress (see Figure 9.3*C*). In the case of major continental strike-slip faults, however, which are controlled by the relative movement of large crustal blocks, the distinction is less clear. Many transform faults appear to originate at abrupt changes in orientation of the severed continental margin (see Figure 15.2) particularly where the margin is nearly parallel to the spreading direction.

One of the best-known examples of a transform fault is the San Andreas fault of California (Figure 14.12). This fault forms the plate bound-

Figure 14.11 Nature of a transform fault (see text).

Figure 14.12 The San Andreas transform fault (see text). Based on Hallam (1973), Fig. 24.

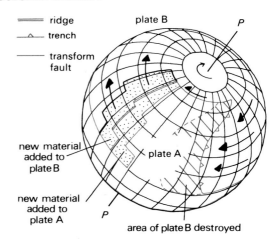

Figure 14.13 Relative movement between two plates expressed as an angular rotation about a pole. The movement of plate B relative to plate A takes place parallel to the small circles about the pole of rotation along bounding transform faults. New material added to both plates at the ridge is balanced by material destroyed by subduction below plate A. Based on Dewey, J.F. (1972) *Scientific American*, May.

ary between the Pacific plate on its west side and the American plate on its east, transforming the divergent motion across the East Pacific ridge to the south to transcurrent motion over a distance of 2800 km until the boundary again becomes a spreading ridge west of Oregon. The direction of this fault thus tells us the direction of relative motion of the Pacific and American plates.

14.4 Geometry of plate motion

Once it is accepted that plates behave as 'rigid' shells, their relative motion across the surface of the globe obeys the simple rules of motion on a sphere. Any relative movement between two plates on the surface of a sphere can be described as an angular rotation about an axis which will intersect the surface of the Earth at two points called the *poles of rotation* for that movement. Figure 14.13 illustrates this principle. The dis-

placement of plate B relative to plate A is an angular rotation about the pole *P*. The direction of movement is parallel to a set of small circles on the globe about the axis *PP*. If the displacement takes place by the opening of an ocean between two bounding transform faults, these faults will also be small circles about the pole of rotation, as they must be parallel to the direction of relative motion. The speed of relative motion may be described as an *angular velocity* which is the speed of rotation about the axis.

The transform fault method was first used to investigate the motion of the Pacific plate relative to the American plate which has a pole situated in the North Atlantic (Figure 14.15). This movement gives apparent velocities that increase southwards away from the pole of rotation. It is important to realize that, although the angular velocity is constant, the *tangential velocity* at the surface varies from a minimum at the pole to a maximum along the great circle at 90° from the pole (see Figure 14.13). By using transform faults and spreading rates, the poles and relative angular velocities of several plate pairs were established. Relative velocities of the remaining plate pairs were then found using the *triple*

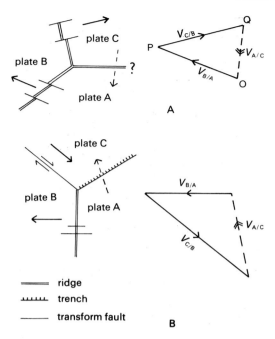

Figure 14.14 Determination of the relative velocities of three plates meeting at a triple junction (see text for explanation).

junction method (Figure 14.14). When three plates meet in a triple junction and the velocities of two of them are known, the velocity of the third can be calculated by drawing a *vector triangle*. In the example shown in Figure 14.14*A*, three plates A, B and C, separated by three spreading ridges, meet at a triple point. The relative velocity of B with respect to A, $V_{B/A}$, is given by the spreading rate on the A/B ridge and the direction of the offsetting transform faults along that ridge. It can be represented by the vector OP, whose length is proportional to the velocity $V_{B/A}$. Similarly, the relative velocity of C with respect to B, $V_{C/B}$, can be represented by the vector PQ. If we suppose that the velocity of A relative to C, $V_{A/C}$, is unknown, it can be calculated by completing the vector triangle by joining QO. Let us assume for simplicity that plate A is stationary. The velocity of plate B is then given by $V_{B/A}$ or OP and the velocity of plate C by $V_{C/A}$ or OQ.

Figure 14.14*B* shows the method applied to a triple junction between a ridge, a trench and a transform fault. The spreading rate at the ridge

Figure 14.15 The major plates showing poles of rotation for six plate pairs and approximate linear velocity vectors relative to the Antarctic plate. 1, America–Africa; 2, America–Pacific; 3, Antarctica–Pacific; 4, America–Eurasia; 5, Africa–India; 6, Antarctica–Africa. After Vine, F.J. and Hess, H.H. (1970) in *The Sea*, vol. 4, John Wiley, New York.

Figure 14.16 The western Indian Ocean region showing the discordance in magnetic anomaly patterns and transform direction at anomaly 5 (33 Ma ago) caused by a change in spreading direction. Newer ocean floor (post anomaly 5); stippled. Older anomalies (23, 25, 30) and transform faults give former relative velocity vectors. Based on Laughton, A.S., McKenzie, D.P. and Sclater, J.G. (1973) in *Implications of Continental Drift to the Earth Sciences*, eds. D.H. Tarling and S.K. Runcorn, Academic Press, New York, Fig. 1.

can be used to determine $V_{B/A}$. The direction of relative motion between B and C is given by the transform fault. The direction of convergence across the trench may also be known from transform faults offsetting the trench. By completing the vector triangle the rate of convergence between A and C can also be found. Thus, working from plate boundaries with known relative motions, a complete picture can be built up of all plate velocities. Actual tangential or linear plate velocities are in the range 2–12 cm yr^{-1} and are illustrated in Figure 14.15 together with the poles of rotation for six major plate pairs. It must be remembered that linear velocities as shown on a map will vary in amount and direction depending on their position in relation to the pole of rotation.

Changes in relative plate motion can be recognized from discordances in the ocean stripe and transform pattern that indicate a change in the position of the pole of rotation. A good example is seen in the Indian Ocean (Figure 14.16) where the northward movement of India relative to the Antarctic plate to the south changed abruptly about 33 Ma ago to a northeasterly movement, causing a new ridge axis to be formed at an angle of 45° to the old direction.

14.5 Driving mechanism for plate motion

Despite the explosion of research into plate tectonics that has taken place in recent years, we seem to be very little closer to an understanding of the fundamental mechanism that drives plate motion. As long ago as 1928, Arthur Holmes suggested *convection currents* in the solid mantle as a mechanism to explain crustal tectonics and continental drift, and it is now generally believed that some kind of convective flow pattern in the mantle provides the driving force for plate motion. However, there is still considerable debate about the nature and pattern of convective circulation, and whether it involves the whole or only part of the mantle.

The ultimate source of energy for tectonic processes is heat. The heterogeneous distribution of the flow of heat leaving the Earth is converted into density imbalances which in turn provide gravitational energy. This can work in two main ways (Figure 14.17). The rise of hotter mantle material below an ocean ridge produces a large low-density bulge which exerts a lateral gravitational pressure on the plates on either side. This

Figure 14.17 Driving mechanism for plate motion (see text).

is the *ridge-push* mechanism. The second mechanism is *slab-pull* which is the gravitational effect of cooler denser material in and around the sinking slab. Another mechanism is *mantle drag* where lateral convective flow within the asthenosphere effectively pulls the plates along. Although all three mechanisms play some part in driving plate motion, calculations of the likely magnitudes of the forces involved suggest that ridge push and slab pull are dominant and that mantle drag is much less important.

Further reading

Continents Adrift and Continents Aground (1972) Readings from *Scientific American*, W.H. Freeman & Co., San Francisco. Contains reprints of important articles on plate tectonics.

Cox, A. and Hart, R.B. (1986) *Plate Tectonics: How it Works*, Blackwell Scientific Publications, Palo Alto, California. Thorough and readable treatment of three-dimensional geometry and kinematics, with examples.

Dewey, J.F. and Bird, J.M. (1970) Mountain belts and the new global tectonics. *J. geophys. Res.* **75**, 2625–2647.

Hallam, A. (1973) *A Revolution in the Earth Sciences*, Clarendon Press, Oxford. A very readable account of the historical development of plate tectonics theory.

Uyeda, S. (1978) *The New View of the Earth: Moving Continents and Oceans*, W.H. Freeman & Co., San Francisco.

Wyllie, P.J. (1976) *The Way the Earth Works: An Introduction to the New Global Geology and its Revolutionary Development*, John Wiley & Sons, New York. An excellent and very readable account of plate tectonics at an introductory level.

15 Geological structure and plate tectonics

In sections 13 and 14 we discussed how the shape of the Earth's surface and the present pattern of tectonic activity can be related to a scheme of plate movements. In this final section we examine the relationship between geological structure and plate movements. The plate tectonic model has been successfully applied to the geological history of the last 200 Ma. A detailed knowledge of plate movements over this period has been built up from our knowledge of oceanic magnetic stratigraphy and continental palaeomagnetism, and can be checked by conventional stratigraphic methods. How far back into the geological past the plate tectonic model can be extended, how-

ever, is a matter of great debate, and one of the important tasks of the structural geologist is to attempt to relate patterns of deformation in old rocks to some scheme of crustal movements.

15.1 Recognition of inactive plate boundaries

The recognition of presently active plate boundaries depends on seismic activity. In interpreting the geological record, we must use other criteria for the recognition of plate boundaries. To some extent, the record over the last 200 Ma since the break-up of the supercontinent Pangaea (see Figure 14.1) can be reconstructed by extrapolation backwards in time, using particularly the oceanic record. However, before that period, there is no oceanic record, since all older oceanic crust will have been destroyed, and we must rely exclusively on the interpretation of continental geology. By examining the way in which structures are related to present-day plate boundaries and plate movements, we can seek to interpret older structures by analogy.

15.2 Structure of constructive boundaries

The continental record of structures associated with recent constructive boundaries is well documented. Evidence is available from the severed margins of the newly formed Atlantic and Indian Oceans and from the incipient constructive boundary represented by the Gulf of Aden–Red Sea rift (Figure 15.1). These rifts meet the great African rift system in a triple junction in the Afar region of Ethiopia. The geological history of this area shows that a domal uplift around 1 km high and 1000 km wide formed during Mesozoic times and was eventually broken by a three-pronged rift associated with deep-seated alkaline vulcanicity. Extensional normal faulting and crustal thinning produced the three rift valley systems, and was accompanied by rather higher-level basaltic volcanics. The final stage of separation is only seen in the Red Sea and Gulf of Aden rifts where thin strips of oceanic crust have formed. The Gulf of Aden has been slowly opening over the last 20 Ma along a continuation of the Carlsberg ridge spreading axis in the

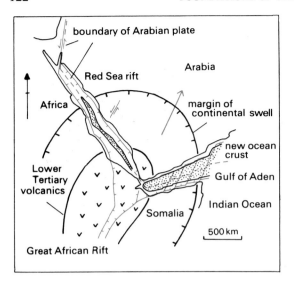

Figure 15.1 Structure of the Red sea–Gulf of Aden–African Rift system. Based on Gass, I.G. (1970) *Phil. Trans. R. Soc. Lond.* **267**A, 369–381.

north–western Indian Ocean (see Figure 14.16).

This triple rift system has been taken as a model for the process of continental break-up leading to the formation of an ocean (Figure 15.2). It has been suggested that the break-up of Pangaea took place by the linking together of a series of jagged rift-fractures of this type situated over mantle '*hot spots*' marked by domal uplifts and volcanism. The third arms of the triple junctions have been called '*failed arms*' since they never develop into oceans but nevertheless exhibit characteristic associations of

structures, sediments and vulcanicity that enable them to be identified in the geological record.

An important feature of some of the rifted margins of Pangaea is the presence of coast-parallel dyke swarms associated with extensive outpourings of tholeiitic basalts, as can be seen for example in eastern Greenland and in the Deccan area of India.

In summary, therefore, we might expect to find the following tectonic features associated with past constructive boundaries— (i) a series of *domal uplifts*; (ii) *extensional faulting* both parallel to the coast-line and defining '*failed-arm*' *graben*; (iii) extensive *vulcanicity* including *extensional dyke swarms*.

15.3 Structure of conservative boundaries

The best documented example of a large *continental transform fault* is the San Andreas fault of California (Figures 15.3, 14.12), where the associated tectonic effects over a period of about 30 Ma have been studied in considerable detail.

A simplified sequence of steps illustrating the plate tectonic history of the region is shown in Figure 15.3*B*. Before about 30 Ma ago, the western border of the American plate was represented by a trench associated with the subduction of the northeast-moving Farallon plate (stage 1). The San Andreas transform fault seems to have commenced when a section of the northwest-moving Pacific plate met the American plate (stage 2). The sense of movement on the American

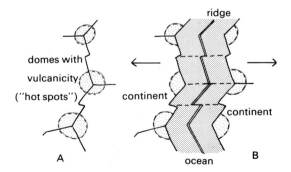

Figure 15.2 Model for the formation of an ocean by continental rifting. After Dewey, J.F. and Burke, K. (1974) *Geology* **2**, 57–60, Fig. 1.

Figure 15.3 Strike-slip fault tectonics illustrated by the San Andreas fault zone in Southern California. *A*. Fault distribution in Southern California with major strike-slip faults in colour. After Anderson, D.L. (1976) 88–102. *B*. Evolution of plate movements in the area of the San Andreas fault, assuming the American plate to be stationary. 1, 53 Ma ago; 2, 30 Ma ago; 3, 10 Ma ago; 4, present. See text. After Atwater, T. (1970) *Bull. geol. Soc. Am.* **81**, 3513–3536. *C*. Detail of strike-slip fault tectonics in the Los Angeles sector of the San Andreas fault zone showing major faults, anticlinal fold trends, and sedimentary basins—see text. After Howell, D.G., Crouch, J.K., Greene, D.S., McCulloch, D.D.S. and Vedder, J.G. (1980) in *Sedimentation in Oblique-Slip Mobile Zones*, eds. P.F. Ballance and H.G. Reading, Spec. Publ. Int. Ass. Sediment. **4**, 43–62, Fig. 10. *D*. Models explaining the origin of compressional structures (folds and thrusts) and extensional structures ("pull-apart" basins) as a result of either fault terminations, or overlapping faults. After Reading (1980) Fig. 3.

plate boundary then changed from convergent to strike-slip along a particular section of the boundary between two transform faults on the Farallon plate. Subduction continued to the north and south. With time the length of the San Andreas fault grew as more of the ridge was subducted.

The effect on the geology of western California was complex (Figure 15.3*A*). A block of early highly deformed trench sediments belonging to the original American plate boundary has now been carried northwards to become the Coast Ranges. Several other large blocks have also been displaced dextrally as part of a complex set of movements on a wide fault zone associated with the San Andreas fault. An interesting feature is the bend in the main fault near Santa Barbara. The geometry demands that dextral strike-slip motion will cause compression across the region of the bend, which accounts for the overthrusting found in this area. The bend is thought to be due to the effect on the American plate boundary of the intersection of the southern transform fault, which would at one time have been continuous with the major Garlock sinistral fault on land. When this transform fault reached the American plate boundary (stage 3 of Figure 15.3*B*) the direction of relative motion would have become convergent for a period until the next section of ridge collided with the continent. At this point, transform motion along the plate boundary would have been resumed, leading to the present situation (stage 4). During this movement the ridge system itself was moving northwards relative to a fixed point (e.g. Los Angeles) on the American plate.

The continental transform fault is actually a zone about 100 km wide between two relatively undeformed blocks. The total strike-slip displacement of the plates on either side has been distributed through this zone, partly as movements on a number of smaller faults roughly parallel to the main fault, and partly in the form of secondary compressional and extensional structures caused by the redistribution of stresses arising out of these fault movements.

Part of the fault zone is shown in Figure 15.3*C*. In this area a number of roughly parallel dextral

strike-slip faults branch from the sinistral Santa Monica fault. Many of these faults terminate or overlap with similar faults. The blocks between the faults contain *extensional 'pull-apart' sedimentary basins* and also *compressional folds* (see Figure 15.3*D* and section 9.13).

This case history illustrates the complexities of structure that may be associated with transform faults, and provides a warning against over-simple explanations of geologically older and less well-documented examples.

15.4 Structure of subduction zones

Seismic investigations have shown that modern *ocean trenches* typically have a V-shaped form with the steeper side towards the direction of dip of the *subduction zone*. Although the floor of the trench is 2–3 km deep, there is usually up to 2 km of sediment fill. These sediments are flat and undeformed along the trench bottom, but show complex folding and thrusting on the inner trench wall where the Benioff zone comes to the surface.

This deformation is clearly related to the subduction process and can be studied more conveniently in zones of rather older trench sediments that have been uplifted above sea level between the present subduction zone and the volcanic arc. Examples of uplifted Mesozoic–Tertiary trench assemblages include the Franciscan

Figure 15.4 Structure of the Makran subduction zone. *A.* Location of the Makran subduction zone in the Gulf of Oman. The subduction zone marks the boundary of the Arabian and Eurasian plates (inset) and is truncated on its eastern side by a transform–strike-slip zone. Stars mark volcanic centres along the active volcanic arc. The trench is obscured by a thick accretionary prism which is partly offshore and partly onshore. *B.* Simplified structural map of the onshore Makran accretionary prism showing major folds and reverse faults. Note lateral facies change between Talar and Parkini formations. *C.* Section across the southern limb of the Kulanch syncline (see *B*) showing N-younging sequences bounded by reverse faults. Pg, Panjgur; Bg, Branguli; Pk, Parkini formations. *D.* Interpretative section across *B* showing suggested structure during the Pliocene. Note that the uplift of the northern limb of the syncline is attributed to a duplex structure. *A–D* after Platt *et al.* (1985).

B

reverse fault
normal fault
15km fault

TALAR shelf sandstones
PARKINI slope mudstones
BRANGULI slope ms & ss
PANJGUR abyssal plain turbidites

KULANCH SYNCLINE

30'

25°15'N

PASNI

30'

GULF OF OMAN

15'

63°E

C

N

S

500 m

Rumra Kaur

0

Pk

Bg

Pg

Bg

Pk

Pg

Bg

Pk

Bg

1km V·H

A

30°N

Sultan

Taftan

LUT

Bazman

HAMUN-I MASHKEL

JAZ MURIAN

MAKRAN

MURRAY RIDGE

5
10
15
20
25
30

67°E

STRAIT OF HORMUZ

OMAN

GULF OF OMAN

5
10
15
20
25
30

OWEN FZ

59°E

Eurasian Plate
Arabian Plate
Indian Plate
African Plate

D

prism front

shelf

slope

abyssal plain

underthrust abyssal plain sequence

oceanic crust

10km V·H

Early Pliocene

'geosyncline' of the Californian coast ranges (see Figure 15.3), parts of the thrust complex of the Banda Arc of Indonesia, and various sections of the Alpine–Himalayan orogenic belt. These examples however are complicated by subsequent tectonic activity—mainly as a result of the collision of continental plates.

The early stages of subduction (see Figure 14.10) appear to be marked by the development of imbricate thrust slices of oceanic crust carrying cherts and shales, sometimes with upper mantle peridotites. These slices pile up to cause crustal thickening on the down-dip side of the subduction zone, which causes gravity sliding of material into the trench. The *olisthostromes* or gravity-slide 'mélanges' formed by this process are a prominent feature of many subduction zones. As the process continues, a thick wedge of clastic sediments (*flysch*) derived from the volcanic belt develops across the arc-trench gap and spreads into the trench, where it becomes involved in the continuing underthrusting. The pile of deformed material that results from this process is termed an *accretionary prism*.

The Makran complex. The accretionary prism of the Makran lies along the continental margin of Iran and Pakistan on the north side of the Gulf of Oman (see Figure 15.4). The complex is formed by the northward subduction of the oceanic part of the Arabian plate beneath the Eurasian plate. The subduction zone is terminated on its eastern side by the Murray transform fault that separates the Arabian and Indian plates (see Figure 14.16) and continues northwards on the continent as the Quetta–Chaman fault system (see Figure 15.5A). The subduction zone ends in the west at the Straits of Hormuz, where the Arabian and Eurasian continents are in contact. A sequence of sediments 6–7 km thick covers oceanic crust in the Gulf of Oman, which is thought to be between 70 and 120 Ma old.

The active volcanic arc consists of a chain of Cenozoic volcanoes situated 400–600 km north of the coast. There is no topographic trench, and the accretionary complex is unusually broad, about 300 km in width, more than half of which lies onshore. Seismic reflection profiles across the

offshore part of the complex show a linear pattern of ridges with intervening troughs. Folding appears to take place initially at the southernmost or frontal part of the prism, which seems to have migrated southwards at a rate of 10 km/Ma. These frontal folds are then incorporated into the accretionary complex by uplift along a basal thrust. Little subsequent deformation appears to have occurred in this sector of the complex. However, 70 km to the north of the present front, a further uplift occurs which rises eventually above sea level 100 km north of the front to form the onshore Makran. Here a thick, faulted flysch sequence is exposed, extending about 200 km inland to the north.

The onshore structure, summarized in Figure 15.4B, affects a concordant sequence of marine sediments commencing with Oligocene to mid-Miocene abyssal plain deposits, followed by upper Miocene slope deposits, and by a late Miocene to Pliocene shallow-water shelf sequence, indicating rapid shoaling of the sedimentary prism in the mid-Miocene. There is apparently no field evidence for accretionary structure prior to the deposition of the slope and shelf sediments. Nor is there evidence for the progressive growth of structures during deposition, although the growth of gentle folds might be undetectable owing to the effects of the later deformation.

The main deformation, which caused 25–30% shortening, occurred after the early Pliocene (4 Ma ago) at a time when the accretionary front probably lay 70–100 km south of the present shore line, and has resulted in a series E–W to ENE–WSW, asymmetric, south-facing folds and associated reverse faults (Figure 15.4C).

The uplift of the onshore Makran and the accompanying deformation are thought to have been accomplished by *underplating* at depth (see Figure 15.4D). It has been suggested that this process may have operated by the formation of a progressively widening duplex at a ramp in the basal thrust. Such a structure could have caused tilting of the upper part of the sequence, leading to shoaling and possibly to syn-sedimentary deformation, before the major folding and faulting.

15.5 Continental collision zones

The collision of two continental plates produces a zone of very complex structure as illustrated by much of the Alpine–Himalayan orogenic belt. The reason for the complexity lies partly in the fact that the structures of two formerly separate continental margins have been brought together, often in a very intimate fashion. The problem is illustrated in Figure 14.10, which demonstrates schematically the effects of telescoping an active continental margin exhibiting subduction tectonics with an opposing passive continental margin. This is the simplest possible situation. The Alpine and Himalayan chains show evidence of several subduction zones and island arcs that have now coalesced.

The Central Asia region. The complexity of the tectonic pattern in Central Asia resulting from the collision between India and Asia is illustrated in Figure 15.5. We shall now discuss some of the reasons for this complexity.

The southern boundary of the Eurasian plate is marked by the *Indus suture* (now a steep fault) which lies on the north side of the Himalayas. At its western end, the suture is terminated by the large Quetta–Chaman strike-slip fault. West of this fault lies the Arabian plate (see Figure 14.16). At the eastern end of the suture the plate boundary runs southwards again along a large dextral strike-slip fault which continues through Burma to connect with the Indonesian subduction zone (see Figure 14.15). This fault also continues into the Indian Ocean as the now largely inactive Ninety-East ridge transform fault. The Indian plate, therefore, may be regarded as a two-pronged wedge that has driven northwards into the Eurasian plate between two large transform faults.

The initial collision appears to have taken place during the Eocene when the leading edge of the Indian continent first made contact with Eurasian continental crust. Prior to this event, the intervening oceanic lithosphere, part of the 'Tethys ocean', was being subducted below the Eurasian plate. Deformed remnants of Tethyan sequences are preserved immediately south of the suture (Figure 15.5C).

It has been estimated that northward movement of India continued for at least a further 1500 km after the initial collision. Part of this movement must have been accommodated by crustal shortening within the Eurasian plate, the effects of which are to be seen in the broad zone of deformation which extends up to 3000 km north of the suture. Structures attributed to this deformation include *compressional fold and thrust belts*, *strike-slip faults*, and *extensional rift systems*. Some of the convergent motion may also have been caused by the closure of a small ocean basin or basins between the Tibetan block and the main Eurasian plate.

The compressional structures are confined to narrow belts separated by wide areas of comparatively undeformed crust. It is estimated that between 750 km and 1000 km of shortening has been achieved in these compressional belts and that the remaining convergence must have been taken up by lateral movements along the major strike-slip faults and by subduction of oceanic crust. In the west, northward movement of the western prong of India would tend to drive large blocks of crust westwards along several northwest–south-east dextral faults. In this way the Eurasian plate appears to have been shortened in a north–south direction and ·elongated in an east–west direction. In the north and northeast also, northwest–southeast extension is indicated by the Baikal rift and Shansi graben systems.

The most spectacular compressional effects of the collision are exhibited in the thrust and fold belt of the Himalayas, south of the Indus suture (Figure 15.5B). Here a significant component of the crustal shortening, estimated at between 300 and 700 km, has taken place by the overlapping of thick crustal slices containing Precambrian basement belonging to the Indian plate. The crust in this area is double its normal thickness and Figure 15.5C shows in simplified form how this doubling of the crust could have been achieved, by underthrusting the main Indian crustal slab along a low-angle thrust plane (the main mantle thrust) outcropping about 130 km south of the suture.

Three stages in this process are shown in

A

B

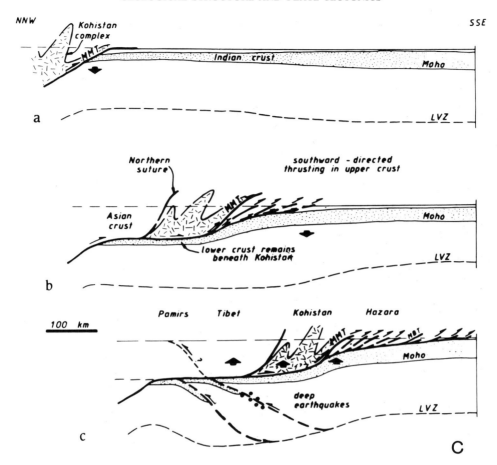

Figure 15.5 Principal tectonic features of Central Asia thought to relate to collision with the Indian continent. *A*. Map showing principal compressional fold belts (wavy lines), thrusts (tooth-edged lines), extensional graben or rifts (lines with tick on downthrow side), and strike-slip faults (lines with arrows indicating sinistral or dextral motion). Coloured arrows indicate compressional or extensional fault-plane solutions from recent earthquakes. After Molnar and Tapponnier (1975), Fig. 4. *B*. Section across the Kohistan region in the Pakistan Himalayas, from the Indus suture to the frontal Himalayan thrust, after Coward, M.P., Jan, M.Q., Rex, D., Tarney, J., Thirlwall, M. and Windley, B.F. (1982) *J. geol. Soc. Lond.* **139**, 299–308, Fig. 7. *C*. Interpretative profiles showing how the structure of *B* can be explained by underthrusting of the Indian plate below itself along the main mantle thrust (MMT). After Coward and Butler (1985), Fig. 3.

cartoon form in Figure 15.5C. The suture marks the position of the former subduction zone. After the Asian and Indian continents made contact along the suture, i.e. when the intervening oceanic lithosphere had been subducted (profile *a*), further convergence between the two plates took place by underthrusting Indian lithosphere beneath the Asian plate, accompanied at shallow levels by southward overthrusting, propagating

from the Main Mantle thrust forwards into the cover of the Indian plate. This shallow overthrusting produced an imbricate thrust stack, or *foreland thrust belt*, detaching on the Himalayan boundary (or frontal) thrust which outcrops 300 km south of the suture (profile *b*). At depth, steep backthrusts may have been developed (see profile *c*) as underthrusting proceeded.

15.6 The Caledonian orogenic belt in Britain

The problems of applying plate tectonics of pre-Mesozoic orogenic belts are well illustrated by the British sector of the Caledonian orogenic belt. Despite (or perhaps because of) the fact that this is probably the most comprehensively studied belt in the world, there is no general agreement on a tectonic interpretation.

We shall take as examples of the structural variation across this belt, four important regions: The *Moine thrust zone* at the northwestern margin, the *Grampian Highlands*, the *Southern Uplands* and the *slate belt* of Wales (Figure 15.6).

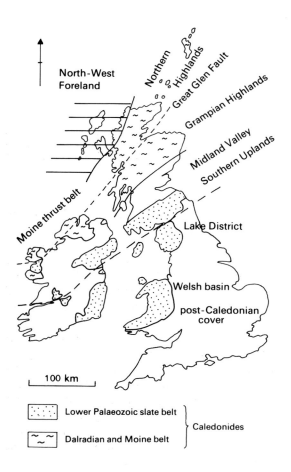

Figure 15.6 Map of the British Isles showing the main tectonic zones of the Caledonian orogenic belt.

The Moine Thrust Zone. This zone marks the northwestern boundary of the Caledonian orogenic belt in Britain (Figure 15.7). It is up to 11 km wide and extends from Loch Eriboll on the north coast of Scotland to Skye, a distance of 190 km. The zone consists of several separate nappes resting on basal thrusts. These nappes die out laterally as the individual thrusts converge. The lowermost nappes are duplexes (see section 9.4), containing highly imbricated sequences of thin Cambro-Ordovician quartzites and limestones, which have been thrust over the unfolded Cambro-Ordovician cover of the foreland, where it rests on Precambrian basement. The middle nappes contain Lewisian (early Precambrian) crystalline cores and exhibit large-scale recumbent folding. The uppermost nappe is composed of the late Precambrian Moine complex with a mylonitized base. The total displacement on the thrust complex has been estimated at up to 100 km.

The evolution of the structure is summarized in Figure 15.7*B*. The thrusting is thought to have developed first in the east with the Moine thrust, which cut upwards and westwards from the basement into the Cambro-Ordovician cover. The sole (or basal) thrust developed next, in part following the base of the Cambrian. Continued movements caused an imbricate zone to form within the cover by the development of steep reverse faults which climb up from the sole thrust. At this stage, the sole thrust would form the floor, and the Moine thrust the roof thrust, of a simple duplex structure containing the imbricate zone. Continued movements then caused the southeastern portion of the early duplex containing the Lewisian basement to climb up over the imbricated Cambro-Ordovician, forming overfolds as it did so to produce the middle nappes (e.g. the Arnaboll nappe—see stage 4). As the lower nappes moved westwards they carried the upper nappes above them in piggyback fashion.

The Moine thrust zone formed in Devonian times during the closing stages of the Caledonian orogeny. It is younger than several other major thrusts and slides which formed during the early Ordovician Grampian orogeny of the Scottish Highlands, and has been attributed to the main

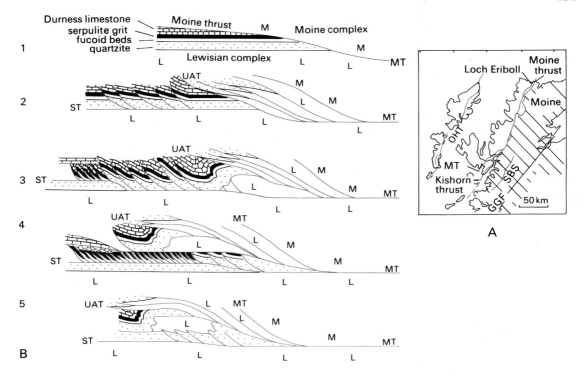

Figure 15.7 Structure of the Moine thrust zone of NW Scotland. *A*. Location map. *B*. Diagrammatic sections across the north end of the thrust belt at Loch Eriboll showing stages in the evolution of the thrust zone. Stage 5 shows only part of the section. See text for explanation. After McClay and Coward (1981), Fig. 7. MT, Moine thrust; OHT, Outer Hebrides thrust; ST, sole thrust; UAT, Upper Arnaboll thrust; SBS, Sgurr Beag slide; GGF, Great Glen fault.

continental collision that occurred further south along the line of the Solway Firth (see Figure 15.10). There is an obvious analogy here with the Central Asia region discussed above (see section 15.5), in the sense that the most recent crustal shortening took place along the Moine thrust zone which is situated nearly 300 km northwest of the collision suture.

The Grampian Highlands. The Scottish Highlands comprises the central metamorphic zone of the Caledonian orogenic belt and contains the most complex structures. Many of the early investigations into structures and structural sequences took place there. The Grampian Highlands is the part which lies south of the Great Glen fault (Figures 15.6, 15.8*A*). The rocks consist of metasediments of the Grampian division, which pass upwards into the late Precambrian to

Cambrian Dalradian supergroup. The major structure consists of large recumbent folds, tens of kilometres in amplitude, that appear to pass downwards into upright folds with much smaller amplitudes in a central steep belt or *root zone* (Figure 15.8*B*). These major folds are associated with *slides* of both thrust and lag type (see section 1.2).

The main deformation, including possibly the first two phases of major folding, took place during the Grampian orogeny in Lower Ordovician times. This primary structure has been ascribed in one interpretation to severe compressive shortening at depth, causing an upwards and outwards flow of material squeezed from a root zone (Figure 15.8*D*). It is suggested that this intense compression resulted from the collision of an island arc, situated in the region now occupied by the Midland Valley of Scotland, which

A

D

B

C

moved towards the Highlands as a result of the subduction of a small oceanic plate along the Highland border. Later strike-slip movements have obscured the original relationships. It is thought that the continental crust was thinned by extension before the deposition of the Dalradian sediments, and that this partly explains the ease with which this zone subsequently became compressed. However, an alternative and more recent interpretation (Figure 15.9A) involves major crustal-scale ductile overthrusting from the south related to the obduction of a large-scale ophiolite nappe (Figure 15.9B stage c).

The major folds and slides are refolded by generally upright northwest–southwest major folds with wavelengths of several kilometres. There appear to be two or perhaps three generations of these folds, with varying orientations, that produce marked interference structures in the outcrop pattern. Each of these sets of major folds is associated with well-developed foliations and minor folds. Further deformation produced several sets of minor structures, mainly crenulation cleavages and kink bands. Altogether eight separate phases of deformation have been recognized in the southwestern Highlands, but many of these are only of local significance. The complexity of the large-scale structure may be illustrated by a map of the Beinn Dorain–Schiehallion area in the central part of the Grampian Highlands (Figure 15.8C). This map shows excellent examples of interference struc-tures (see section 10.4) attributed to the super-imposition of four generations of major folds.

The later upright folds are ascribed to the compressional effects of collision, firstly with a volcanic arc terrane (Figure 15.9B stage d), and secondly with the main European plate in the late Silurian to Devonian period (Figure 15.9B stage e).

The Southern Uplands belt. This belt (Figure 15.10A) has been interpreted as a Lower Palaeozoic example of an *accretionary prism* or *subduction complex*, comprising a set of thrusts and related folds resulting from the under-thrusting of an oceanic slab at a trench. The combination of thrusts and related folds formed in such a tectonic regime has been termed a *synthetic thrust belt*. The area consists of a 60 km-wide zone of folded and faulted Lower Ordovician to Upper Silurian sediments and minor volcanics forming part of the Caledonian orogenic belt (see Figure 15.6). The basal part of the sequence consists of oceanic basalts and cherts overlain by a very thick pile of greywackes and shales. The belt consists of at least ten separate fault blocks each containing sequences which young towards the northwest, but the belt as a whole becomes progressively younger towards the southeast. This arrangement is thought to reflect underthrusting related to a trench situated in the Solway Firth region in Ordovician to Silurian times (Figure 15.10B, C).

Some of the fault blocks are thought to have undergone considerable strike-slip displacement in relation to adjoining blocks, and the belt as a whole is regarded as a *displaced terrane* in relation both to the Midland Valley to the north and to the Lake district to the south.

The slices with their bounding faults are believed to have been steepened by compression (Figure 15.10D) due partly to the continued subduction and partly to subsequent continental collision. The folds within the slices are asymmetric with long north-west-facing limbs and short south-east-facing limbs, often partly thrust-out. The deformation is most intense in the ductile Moffat shale formation which may have formed a décollement or sliding horizon

Figure 15.8 Structure of the Grampian Highlands of Scotland. *A*. Location map showing area of map *C* and line of profile *B*. After Johnstone, G.S. (1966) *British Regional Geology—The Grampian Highlands*, HMSO, Fig. 3. *B*. Structural block diagram across the Grampian Highlands to show the geometry of the major structures. AS, Appin syncline (*F1*); BA, Bohespic antiform (*F3*); BLA, Beinn na Lap antiform (*F2*); DD, Drumochter dome (*F1/F2*); KA, Kinlochleven anticline (*F1*); SBS, Stob Bhan synform (*F2*); SMS, Stob Mhor synform (*F2*). After Thomas (1979), Fig. 5. *C*. Map showing complex interference structures between *F1*, *F2* and *F4* between Beinn Dorain and Loch Tummel (see *A*). After Roberts, J.L. and Treagus, J.E. (1979) in *The Caledonides of the British Isles—Reviewed*, eds. A.L. Harris, C.H. Holland and B.E. Leake, *Spec. Publ. geol. Soc. Lond.* **8**, 199–211, Fig. 1. *D*. Simplified diagrams of the primary deformation to explain (*a*) generation of nappes and slides during *D1*, and (*b*) modification of *D1* nappes by *D2* major folds. After Thomas (1979), Fig. 6.

Figure 15.9 *A*. Interpretative profile through the Scottish highlands showing the main Grampian structures as NW-facing and the SE-facing folds (e.g. the Tay nappe) as backfolds (compare Fig. 15.8*D*). MT, Moine thrust; SBS, Sgurr Beag slide; FWS, Fort William slide; BAS, Ballachulish slide; IBS, Iltay boundary slide; GGF, Great Glen fault; HBFZ, Highland boundary fault zone; SUFZ, Southern Uplands fault zone. *B*. Sequential cartoons illustrating an evolutionary model for the Grampian orogeny; *a–c*, an ophiolite is overthrust onto oceanic lithosphere at a fracture zone and thereafter progressively obducted onto the continental rise and shelf, producing the *D*1-2 deformations in the Grampian Highlands; *d*, shortening and thickening of the sedimentary pile leads eventually to a reversal of subduction direction. From Dewey and Shackleton (1984), Figs. 2, 3.

(cf. section 9.4) for the underthrusting. The arrangement of the structures bears some superficial similarities to the onshore Makran complex (see section 15.4 and Figure 15.4).

The Welsh Slate belt. The Welsh fold belt (Figure 15.11) affects a basin containing Cambrian to Silurian sediments and volcanics underlain by Precambrian basement, which outcrops in the northwest in Anglesey and to the southeast in the Welsh borders and in Pembrokeshire. The main deformation is end-Silurian in age although it was preceded by periods of local flexuring that produced several unconformities.

The deformation has resulted in an early low-angle cleavage, seen on the north side of the Harlech dome, that is associated with the Tremadoc thrust belt. Later upright folds associated with a steep cleavage are widespread and locally deform the low-angle cleavage. These upright folds are often rather open structures and only locally become isoclinal or overturned. High strains have been measured in the Cambrian slate belt of North Wales (67% flattening and 157% vertical extension) but other areas (e.g. in the southeast) show very low strains (Figure 15.11*B*).

Compared with the Scottish Highlands and the Southern Uplands, the deformation is much less intense and is also much more variable. The fold trend locally swings from north–south to

Figure 15.10 Structure of the Southern Uplands of Scotland—a possible example of Lower Palaeozoic subduction tectonics. *A*. Simplified geological map of the Southern Uplands showing the principal faults. After Leggett *et al.* (1979), Fig. 1. *B*. Diagrammatic profiles across the area in late Arenig and early Llandovery times, interpreted as an evolving trench. After Dewey, J.F. (1971) *Scot. J. Geol.* **7**, 219–240, Fig. 5. *C*, Diagrammatic profile across the area to illustrate the underthrusting model. Each slice has been detached along the Moffat shale horizon (dotted). After Leggett *et al.* (1979), Fig. 6. *D*. Diagram showing how the orientation of asymmetric folds and thrusts may be changed by steepening due to compression. Moffat shales, ruled. After Eales, M.H., in *The Caledonides of the British Isles—Reviewed*, eds. A.L. Harris, C.H. Holland and B.E. Leake (1979) Spec. Publ. geol. Soc. Lond. **8**, 269–273, Fig. 2.

Figure 15.11 Main structural features of the Welsh fold belt. After Coward and Siddans (1979), Fig. 1. *A.* Tectonic map showing folds, cleavage planes and thrusts, together with areas of older basement. *B.* Diagrammatic structural profile showing the mainly gentle folding and also the tighter folds and thrusts of Carmel Head, Tremadoc and the Berwyns. This interpretation suggests that the thrust may 'bottom out' along a basal décollement plane (see text).

east–west in orientation. One interpretation suggests that the two major thrust zones 'bottom out' into a major décollement horizon along which the sedimentary cover has been shortened and thickened by a push towards the south (Figure 15.9*B*). The deformation appears to die out gradually towards the Midlands craton although several major belts of strike-slip deformation have been recognized along the southeastern margin of the belt.

It is considered likely that a southeasterly-dipping subduction zone lay to the northwest of Wales during the Lower Palaeozoic, but the main deformational event is attributed to end-Silurian plate collision.

15.7 Orogenies in the Precambrian

The problems of reconstruction and interpretation become more acute when we consider the Precambrian orogenic belts. There has been much debate among geologists as to how far back in time the plate tectonic model can be extended or how it should be modified to explain the earlier history of the Earth. Some geologists believe that lithosphere subduction did not commence until later Precambrian times, and others that greater heat loss in the earlier Precambrian was accomplished by the presence of many small oceanic basins, with more rapid subduction producing a higher rate of circulation of oceanic crust.

When we examine the nature of Precambrian orogenic belts we find certain differences which set them apart from their more modern counterparts. While many belts show evidence of subduction-related magmatism, and ophiolite complexes marking collision sutures, as shown by the Alps or the Himalayas, for example, others reveal no evidence of plate collision by the removal of an intervening ocean. Palaeomagnetic evidence and matching of structure on either side of certain belts indicate that the cratons on either side have not been displaced by more than a few hundred kilometres.

Some belts are interpreted as major *shear zones* where displacement has been strike-slip with little or no contraction across the belt. In the later Archaean and Lower Proterozoic, major shear zones are very common and suggest that much of the continental lithosphere, although able to transmit a uniform stress field, was in general more 'deformable' than is the case today. We shall consider three examples of Precambrian tectonics: the Lower Proterozoic Labrador fold belt of Canada, the Nagssugtoqidian mobile belt of Greenland, and the Archaean deformation pattern of the Superior Province of Canada.

The Labrador fold belt. This 400 km-wide belt is part of a continuous Early Proterozoic

Figure 15.12 Precambrian provinces of the Canadian shield. The map shows the Archaean cratons (dotted) and Proterozoic mobile belts of the Canadian shield. *A*, area of Fig. 15.15*A*; *B*, area of Fig. 15.15*B*. SUP, Superior province; NAC, North Atlantic craton; NAG, Nagssugtoqidian.

(c. 1800 Ma) orogenic zone surrounding the Archaean craton of the Superior Province (Figure 15.12). On its northeast side, it is bounded by the North Atlantic craton which consists of the Archaean areas of eastern Labrador and southern Greenland. The belt consists of two parts, a western highly-deformed sequence of volcanics and sediments of Early Proterozoic age (the Labrador 'geosyncline'), and an eastern part consisting of re-worked Archaean basement penetrated by granites. The deposits thicken towards the centre of the trough where they reach about 10 km in thickness. The structure is illustrated in Figure 15.13. The folding and associated thrusting are strongly asymmetric, directed towards the southwest.

This rather deep basin seems to have formed on continental crust as a result of major crustal extension, but there is no evidence that any oceanic crust was formed. It has been suggested that the Superior craton was moved southeastwards in relation to the blocks to the north and east, forming a set of extensional and transtensional zones. These subsequently became the sites of compressional and transpressional deformation, forming the Labrador and other related fold belts, when the movement was

Figure 15.13 A structural interpretation of the Labrador fold belt of Canada. After Dimroth (1980). *A*. Generalized geological section showing main supracrustal rock units and the asymmetric arrangement of folds and thrusts. After Dimroth (1980), Fig. 13.3. *B*. Crustal delamination model for the evolution of the fold belt (see text). After Dimroth (1980), Fig. 13.7.

reversed as a result of a postulated plate collision to the southeast.

The Nagssugtoqidian belt. The Early Proterozoic Nagssugtoqidian belt of Greenland (Figures 15.12, 15.14) is a 300 km-wide zone made up mainly of re-worked Archaean gneisses but including some small patches of post-Archaean metasediments and volcanics (Figure 15.14*A*). The zone is bounded to the south by the Archaean craton and to the north by the Rinkian mobile belt whose tectonic relationship with the Nagssugtoqidian is obscure. In the northeast, the belt is bounded by Archaean rocks.

The southern margin of the Nagssugtoqidian is a broad ductile shear zone (the Ikertoq shear zone) in which granulite-facies Archaean rocks like those south of the margin have been transformed into intensely-deformed amphibolite-facies gneisses. An early phase of strike-slip displacement on the shear zone was followed by the intrusion of a set of dykes whose intrusion pattern has been partly controlled by the shear zone, and then by a later phase of more intense over-thrust displacement towards the south. This shear zone, which is about 50 km wide, grades northwards into a zone of rather less intense deformation which is followed northwards by another, narrower shear zone at Nordre Strømfjord. The latter belt is about 15 km wide and exhibits sinistral transcurrent displacement (Figure 15.14*B*). The map shows very clearly how the structures and rock units north and south of the shear zone bend into parallelism with the shear zone, with individual rock bodies becoming greatly extended and thinned in the process.

These belts of intense ductile deformation exhibit very high strains. Most of the fold axes are sub-parallel to a well-developed lineation that defines the *X* strain axis and is sub-parallel to the direction of simple shear. Folds are typically similar in form, and boudinage of competent layers is common. Original discordances have been virtually eliminated and a marked parallelism of all previous structures, whatever their orientation, is characteristic. These features are all typical of high degrees of homogeneous strain produced by simple shear (see sections 8.8 and 9.5).

There is no evidence within the Nagssugtoqidian belt of plate collision nor any sign of oceanic crust. There are few granites, little vulcanicity and indeed none of the signs of a typical Phanerozoic orogenic belt. The belt appears to represent deep-seated intra-plate ductile deformation associated with both strike-slip and over-thrust displacements. The consistency in pattern and sense of movement on a number of shear zones of Early Proterozic age in Greenland and northwestern Scotland (Figure 15.14*C*) suggests a simple NNW–SSE compression acting across a large Archaean plate.

Archaean tectonics in the western Superior Province. It is generally impossible to subdivide the Archaean crust into mobile belts and cratons, since evidence of mobility is so widespread. The Superior Province of Canada (Figure 15.12) is an example of the *granite-greenstone* type of Archaean terrain where belts or irregular outcrops of volcanic and sedimentary rock (the 'greenstone

Figure 15.14 Main structural features of the Nagssugtoqidian mobile belt of West Greenland. *A*. Map of the Nagssugtoqidian mobile belt in West Greenland showing structural trends and main rock units. After Escher *et al.* (1976), Fig. 71. *B*. Map of the Nordre Strømfjord shear zone in the central part of the Nagssugtoqidian belt (see *A*). Note how the sinistral shear zone is defined by the various lithological units and structural trends becoming aligned in a NE–SW belt in a zone of intense deformation. Black, amphibolites; ruled, metasediments; crosses, granitic and charnockitic intrusions. The blank areas are granitic to tonalitic gneisses. After Olesen, N. Ø., Korstgård, J.A. and Sørensen, K. (1979) *Grønlands geologiske undersøgelse* **89**, 19–22, Fig. 1. *C*. Pre-Mesozoic reconstruction of the North Atlantic region showing major Early Proterozoic shear zones in Greenland and NW Scotland. In W Greenland, the NNW–SSE zones are sinistral, the WNW–ESE zones are dextral and the WSW–ENE zones are overthrust. These senses of displacement are consistent with a regional NNW–SSE compression. After Watterson, J. (1978) *Nature* **272**, 636–640, Fig. 1.

A

B

Figure 15.15 Granite-greenstone tectonics of the Archaean Superior Province, Canada, *A*. Granite-greenstone relationships in the Kenora area. Note the more equidimensional shape of the granite plutons (dotted) and the variable trend of the folds (lines) which in some places 'wrap around' the pluton margins. After Goodwin (1972) *Variations in Tectonic Styles in Canada*, Fig. 6. *B*. Structure of the Bigstone Lake greenstone belt, NW Superior province, Canada. The early deformation (*F*1, *S*1) appears to be related in part to the margins of the granite plutons (dotted) whereas the later (*F*2) is related to major sinistral and dextral shear zones. After Park (1982), Fig. 5.

belts') alternate with large areas of granite or granitic gneiss. The larger greenstone areas often form basins with volcanic and sedimentary successions several tens of kilometres in thickness.

The greenstone outcrops are normally surrounded by intrusive granites, but in some cases there is evidence that the greenstones rest on older granitic basement. It is thought likely that many of the granite areas represent *mantled gneiss domes* (see section 12.4) where the granite has been emplaced by diapirism in the solid state. The earliest deformation seen in the greenstone belts is variable in orientation and often follows the arcuate margins of the granites, indicating

that it is related to the granite emplacement. Later deformation is more regular and is often related to shear zones which cut across both granites and greenstones. Figure 15.15*A* shows a granite-greenstone terrain in the Kenora area of the southwestern Superior Province that is comparatively unaffected by later deformation. Here the granites are approximately equidimensional in shape, and the early fold axes in the greenstone areas have a variable trend, often curving around the granite contacts. This is the type of pattern that would be expected if the granites were emplaced as solid diapirs.

Figure 15.15*B* shows an example of the deformation pattern in the Bigstone Lake greenstone belt of northern Manitoba. The early deformation has produced a tight upright syncline within the greenstone sequence and a foliation that closely follows the margins of the granites. This structure is bent and refolded by structures related to two ductile shear zones that cross the belt, an earlier northeast–southwest zone with sinistral displacement, and a later northwest–southeast zone with a dextral displacement. There appears to be a conjugate set of such zones in this region, indicating a late Archaean north–south compression.

It would appear, therefore, that in this region, an early stage of general crustal mobility with the widespread formation of volcanic basins, the diapiric emplacement of a granitic substratum, and the intrusion of granitic plutons, may have given way to a period when the crust was rigid enough to transmit stress and to form regular sets of shear zones (similar to the Greenland examples previously mentioned), suggesting the commencement of more 'plate-like' behaviour.

Further reading

Anderson, D.L. (1972) 'The San Andreas fault', in *Continents Adrift and Continents Aground*, Readings from *Scientific American*, W.H. Freeman, San Francisco.

Coward, M.P. and Butler, R.W.H. (1985) Thrust tectonics and the deep structure of the Pakistan Himalaya. *Geology* **13**, 417–420.

Coward, M.P. and Siddans, A.W.B. (1979) 'The tectonic evolution of the Welsh Caledonides', in *The Caledonides of the British Isles—Reviewed*, eds. A.L. Harris, C.H. Holland

and B.E. Leake, Spec. Publ. geol. Soc. Lond. **8**, 187–198.

Dewey, J.F. and Burke, K. (1974) Hot spots and continental break-up: implications for collisional orogeny. *Geology* **2**, 57–60.

Dewey, J.F. and Shackleton, R.M. (1984) A model for the evolution of the Grampian tract in the early Caledonides and Appalachians. *Nature*, London **312**, 115–121.

Dimroth, E. (1980) 'Labrador geosyncline: type example of early Proterozoic cratonic reactivation', in *Precambrian Plate Tectonics*, ed. A. Kroner, Elsevier, Amsterdam, 331–352.

Elliott, D. and Johnson, M.R.W. (1980) Structural evolution in the northern part of the Moine thrust belt, NW. Scotland. *Trans. R. Soc. Edin. Earth Sciences* **71**, 69–96.

Escher, A., Sørenson, K. and Zeck, H.P. (1976) 'Nagssug-toqidian mobile belt in West Greenland', in *Geology of Greenland*, eds. A. Escher and W.S. Watt, Geol. Surv. Greenland.

Gass, I.G. (1970) The evolution of volcanism in the junction area of the Red Sea, Gulf of Aden and Ethiopian rifts. *Phil. Trans. R. Soc. Lond.* **267A**, 369–381.

Goodwin, A.M. (1972) 'The Superior Province', in *Variations in Tectonic Styles in Canada*, eds. R.A. Price and R.J.W. Douglas, Spec. Paper geol. Ass. Canada No. 11, 528–623.

Leggett, J.K., McKerrow, W.S. and Eales, M.H. (1979) The Southern Uplands of Scotland: a Lower Palaeozoic ac-cretionary prism. *J. geol. Soc. Lond.* **136**, 755–770.

McClay, K.R. and Coward, M.P. (1981) 'The Moine thrust zone: an overview', in *Thrust and Nappe Tectonics*, eds.

K.R. McClay and N.J. Price, Spec. Publ. geol. Soc. Lond. **9**, 241–260.

Molnar, P. and Tapponnier, P. (1975) Cenozoic tectonics of Asia: effects of a continental collision. *Science* **189**, 419–426.

Park, R.G. (1982) Archaean tectonics. *Geol. Rundsch.* **71**, 22–37.

Park, R.G. (1988) *Geological Structures and Moving Plates*, Blackie, Glasgow and London. An account of how geolog-ical structures may be explained by plate tectonics, looking particularly at the structures of divergent, convergent and strike-slip plate boundaries.

Platt, J.P., Leggett, J.K., Young, J., Raza, H. and Alam, S. (1985) Large-scale sediment underplating in the Makran accretionary prism, southwest Pakistan. *Geology* **13**, 507–511.

Reading, H.G. (1980) 'Characteristics and recognition of strike-slip fault systems', in *Sedimentation in Oblique-Slip Mobile Zones*, eds. P.F. Ballance and H.G. Reading, Spec. Publ. Int. Ass. Sediment. **4**, 7–26.

Roberts, J.L. and Treagus, J.E. (1977) Polyphase generation of nappe structures in the Dalradian rocks of the southwest Highlands of Scotland. *Scott. J. Geol.* **13**, 237–254.

Thomas, P.R. (1979) 'New evidence for a Central Highland root zone', in *The Caledonides of the British Isles—Reviewed*, eds. A.L. Harris, C.H. Holland and B.E. Leake, Spec. Publ. geol. Soc. Lond. **8**, 205–211.

Windley, B.F. (1976) *The Evolving Continents*, John Wiley, New York. Discusses orogeny and the possible application of plate tectonics throughout geological times from the Archaean to the present.

Appendix

Stereographic projection is a graphical method of portraying three-dimensional geometrical data in two dimensions, and of solving three-dimensional geometrical problems. In geology, the method is used mainly for solving problems involving the orientations of lines and planes in crystallography and structural geology. Such problems involve the angular relationship between lines or planes rather than their spatial relationships. Planes plot as lines (usually curved) and lines as points on such a projection.

The orientation of a plane (e.g. bedding, foliation or crystal face) is represented by imagining the plane passing through the centre of a sphere of radius R, the *projection sphere*. The plane intersects the sphere in a circle with radius R called a *great circle*. The two-dimensional projection of the plane is produced on a horizontal plane through the centre O of the projection sphere (see Figure A.1A). In structural geology, the lower hemisphere only is used for projection, whereas in crystallography, the upper hemisphere is used. In the following account, the lower-hemisphere projection is described. Upper-hemisphere projections will be mirror-images across a vertical plane of symmetry. For lower-hemisphere projection, each point on the lower half of the great circle is connected to the point T where the vertical line through O cuts the top of the projection sphere, producing a set of lines TA, TB, etc. that intersect the projection plane in an arc known as the *cyclographic trace* of the great circle. The cyclographic trace of the horizontal plane (i.e. the projection plane) is called the *primitive circle*. The cyclographic trace of a vertical plane is a straight line through O. If the primitive circle is given geographic co-ordinates, then the orientation of any specified plane may be represented. Thus Figure A.1B shows the cyclographic traces of a set of great circles with a N–S strike dipping at 10° intervals from 10° W through the vertical to 10° E.

In practice, stereographic projection is carried out by means of a protractor termed a *stereographic net* or *Wulff net* (Figure A.2A) which

gives the cyclographic traces of the complete set of great circles at 2° intervals about a common axis XX′. The net also gives the cyclographic traces of a family of circular cones about the same axis XX′, again at 2° intervals. These, termed *small circles*, are used to graduate angular distances along the great circles. Equal angles on the surface of the sphere project as equal distances on the net. Thus the Wulff net is often termed the *equal-angle net* in contrast to the *Schmidt*, or *equal-area, net* described below.

The procedure for plotting geometric data is as follows: a piece of tracing paper is laid over the stereographic net and fixed with a pin through the centre O. The primitive circle is marked, and

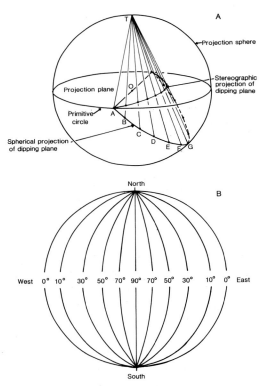

Figure A.1 *A*, Geometry of stereographic projection. *B*. Stereographic projection of planes dipping due east or due west at the various indicated angles. *A*, *B* after Hobbs *et al.* (1986).

142

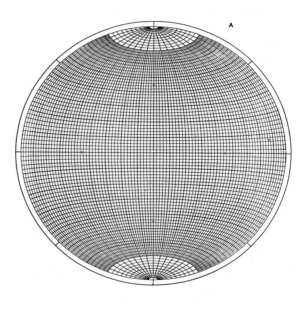

157°), the position corresponding to 157° is marked, and the net rotated until the point situated 90° clockwise of X corresponds with 157°. The appropriate great circle trace is found as before. It should be noted that planes dipping to the east plot in the right side of the stereogram and vice versa.

To plot the position of a line with a plunge of 45° on a bearing of 111°, say, a similar procedure is followed. The bearing 111° is found as before. Since a plunge is measured in a vertical plane, one of the two vertical planes (represented by straight

Figure A.2 *A*, Wulff or equi-angular net. *B*, Stereographic projection of a plane with a strike of 067° dipping towards 157°. A lineation, *L*, pitches 60° NE in this plane. The plunge of the lineation is 45° towards 111°. After Hobbs *et al.* (1986).

given geographical co-ordinates. The plot produced on the tracing paper is termed a *stereogram*. To plot the position of a plane with strike bearing (azimuth) 067°, the position of 067° is marked on the primitive circle, by counting clockwise from N, and the stereogram rotated over the net until 067° corresponds with the axis XX′ of the net. If the plane has a dip of 56° to the SE, the appropriate great circle trace is found by counting 56° inwards from the perimeter (primitive circle) towards the centre (Figure A.2B). Alternatively, if the dip azimuth is given (56° to

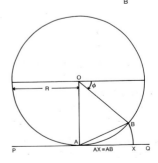

Figure A.3 *A*, Schmidt net. *B*, The principle of the equal area projection. *O* is the centre of the projection sphere, radius *R*. *OB* is the trace of a plane dipping at angle ϕ. *X* is a point on the equal area projection of this plane, *PQ*. After Hobbs *et al.* (1986).

lines on the net) must be rotated to correspond with the position of 111° on the stereogram. The plunge angle is counted inwards from the perimeter to a point situated 45° along the line towards the centre (Figure A.2B).

Angles between any two lines may be found by placing their projected points on a common great circle trace. Angles between any two planes may be found by plotting the *poles* (normals) to the two planes and obtaining the angle between the poles. The pole to a plane is the line (point) situated 90° from the centre of the great circle trace, measured along the vertical trace. Many geometrical problems in structural geology can be solved using various combinations of these simple procedures. For example the *plunge* of a *fold axis* may be obtained by plotting the intersection of great circles representing two planar *fold limbs*, or representing various positions on the fold surface. Poles to a *cylindroidal fold* surface plot on a great circle trace, the pole of which represents the fold axis.

In the statistical analysis of the distribution of structural elements, it is more important to be able to compare the density of lines (points) in different areas of the projection sphere than to compare angles between lines. For the former

purpose, a different type of projection is used, in which equal areas on the projection sphere plot as equal areas on the stereogram. The net used for this purpose is termed an equal-area or Schmidt net (Figure A.3A). The method of projection is illustrated in Figure A.3B. Equal angles on the projection sphere are not projected as equal angles on the stereogram, and the cyclographic traces of the great circles are fourth-order quadric curves. The Schmidt net may be used to plot planes and lines, and measure their angular relationship in exactly the same way as the Wulff net, and is often employed by structural geologists in preference to the latter.

For details of procedures in plotting and analysing density distributions of structural or fabric data, see Turner & Weiss (1963) and Phillips (1971).

Further reading

Hobbs, B.E., Means, W.D. & Williams, P.F. (1986) *An outline of structural geology*, 2nd edn, John Wiley, New York.

Phillips, F.C. 1971: *The use of stereographic projection in structural geology*. Edward Arnold, London.

Turner, F.C. & Weiss, L.E. 1963. *Structural analysis of metamorphic tectonites*. McGraw-Hill, New York.

Index

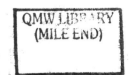